Organometallics 2

Complexes with transition metal–carbon π-bonds

Manfred Bochmann

University of East Anglia

OXFORD
UNIVERSITY PRESS

OXFORD

UNIVERSITY PRESS

Great Clarendon Street, Oxford OX2 6DP

Oxford University Press is a department of the University of Oxford.
It furthers the University's objective of excellence in research, scholarship,
and education by publishing worldwide in

Oxford New York

Auckland Bangkok Buenos Aires Cape Town Chennai
Dar es Salaam Delhi Hong Kong Istanbul Karachi Kolkata
Kuala Lumpur Madrid Melbourne Mexico City Mumbai Nairobi
São Paulo Shanghai Taipei Tokyo Toronto

Oxford is a registered trade mark of Oxford University Press
in the UK and in certain other countries

Published in the United States
by Oxford University Press Inc., New York

© Manfred Bochmann, 1994

First published 1994

Reprinted 1994, 2003, 2004, 2005, 2006, 2007

British Library Cataloguing in Publication Data

Data available

Library of Congress Cataloging in Publication Data
Bochmann, Manfred.
Organometallics 2: complexes with transition metal-carbon
π-bonds/Manfred Bochmann.
(Oxford chemistry primers; 12–13)
Includes bibliographical references and indexes.
1. Organotransition metal compounds. I. Title. II. Series
QD411.8.T73B63 1993 547'.056—dc20 93–22883

ISBN 978-0-19-855813-2

12

Printed and bound in Great Britain by
Antony Rowe Ltd, Chippenham, Wiltshire

Series Editor's Foreword

Organometallic chemistry has been one of the most exciting growth areas of inorganic chemistry over the last 40 years. As well as being an established academic discipline it is of key importance in the design of industrial homogeneous catalysts, and also of new organic synthetic methods.

Oxford Chemistry Primers are designed to give a concise introduction to all chemistry students by providing the material that would usually be covered in an 8–10 lecture course. As well as providing up-to-date information, this series will provide explanations and rationales that form the framework of an understanding of inorganic chemistry. In the second of his books, Manfred Bochmann provides the reader with a clear, timely description of the chemistry of π-bonded carbon donor ligands in transition metal complexes, utilizing well-accepted fundamentals.

John Evans
Department of Chemistry, University of Southampton

Preface

Few areas of chemistry have developed so rapidly, and contributed so widely, not only to the fundamental understanding of the nature of the chemical bond but also to its commercial application, as the organometallic chemistry of the transition metals. Central to this development was the discovery of ferrocene and the recognition of the sandwich structure. This volume highlights the key structural types of metal complexes with unsaturated organic ligands: complexes with alkenes, alkynes, cyclopentadienyls, allyls and arenes. It is the second part of a brief introduction into the principles of organometallic chemistry and complements *Organometallics 1*, which dealt with complexes containing η^1-bonded carbon ligands, such as carbonyls, alkyls, alkylidenes and alkylidynes.

The sequence of chapters reflects the importance of the bonding concepts they introduce. The coordination of alkenes and the closely related alkynes are described first, followed by a discussion of complexes with sandwich structure, such as cyclopentadienyl, allyl and arene complexes. In each chapter, examples of the most important synthetic methods are followed by a discussion of the metal–ligand bonding involved, and an exploration of the reactivity patterns. As in *Organometallics 1*, emphasis has been placed on illustrating the applications of the principles to synthesis or catalysis, often in the form of highlighted excursions. A brief bibliography refers the reader to topical reviews and more comprehensive texts.

M.B.

Norwich
January 1993

Contents

1 Introduction

Broadly speaking, organic molecules or radicals can form two types of complexes with transition metals, those containing metal–carbon σ-bonds, and those interacting with the metal via their π-electrons. The first category was discussed in *Organometalics 1*, the first part of this brief introduction into organometallic chemistry, and was concerned with metal carbonyls and metal alkyl, alkylidene and alkylidyne complexes. Here we will be concerned with π-complexes, i.e. the complexes formed between metals and alkenes, alkynes, alkenyls and arenes.

It is useful to restate some of the more important bonding principles and rules concerning these classes of compounds, although the reader is referred to *Organometallics 1* for a more detailed discussion.

In this context, 'organometallic' complexes are defined as those containing direct metal–carbon bonds. This may occur, as in the case of main group elements, by forming distinct metal–carbon σ-bonds to give metal alkyls, such as $Ti(CH_2SiMe_3)_4$, or by the interaction of metals with the π systems of unsaturated organic molecules. This form of metal–carbon interaction is unique to transition metals (and, to a more limited extent, lanthanides and actinides), i.e. metals whose energetically accessible d-orbitals allow efficient overlap with the π-orbitals of C–C and C–X multiple bonds to give 'π-complexes'. The moiety containing the π-bond may be neutral, such as ethene or ethyne, or formally anionic, as is the case with cyclopentadienyl or cyclo-octatetraenyl ligands.

It has proved useful to determine the number of valence electrons (VE) of a given complex in order to determine whether the complex is likely to be stable or not. It was shown in the case of metal carbonyls (*Organometallics 1*, Chapter 2) that most organometallic complexes obey the 18-electron rule:

18-Electron rule: a stable complex is obtained when the sum of metal d-electrons, of electrons donated from the ligands, and of the overall charge of the complex equals 18.

18-electron rule

When this is the case, the metal acquires the electron configuration of the next highest noble gas. The iron centre in $Fe(CO)_5$, for example, possesses a krypton-like electron shell, whereas a single iron atom has only eight d-electrons.

For transition metals the *18-electron rule* has a predictive function similar to the *octet rule* for main group elements. It provides a simple 'rule-of-thumb' basis for the discussion of structure and bonding, and although there are numerous exceptions, compounds which fulfil this rule tend to represent relative energy minima.

NOTE: Although textbooks will frequently give the electron configuration of a zerovalent transition metal such as Ti as [Ar]$3d^2s^2$ and of Fe as [Ar]$3d^6 4s^2$, in a chemical environment the $4s$ level is **always** higher in energy than 3d: Ti(0) has the configuration [Ar]$3d^4$, Fe is [Ar]$3d^8$ (F. L. Pilar, 1978).

Electron counting and formal oxidation states

In order to determine whether a compound obeys the 18-electron rule or not, it is useful to follow certain electron counting conventions. **Considerations of the formal oxidation state of the metal centre are avoided by treating the metal as zerovalent** (for counting purposes). This means that, irrespective of the physical reality, a chromium atom, for example, is always considered as possessing 6 electrons (d^6), iron as d^8, rhodium as d^9, or platinum as d^{10}.

Considering the metal as formally zerovalent implies that the ligands, too, have to be considered as electroneutral entities. Chloride or methyl ligands are therefore treated as chloride or methyl radicals which donate one electron to the metal. Ligands such as ethene are, of course, neutral and donate two π-electrons to the metal. Similarly, the cyclopentadienyl (C_5H_5) ligand is considered as a pentaradical rather than a $C_5H_5^-$ anion, donating five electrons.

$C_5H_5^-$ 6e⁻		$C_5H_5^{\cdot}$ 5 e⁻
Fe^{2+} d^6		Fe^0 d^8
$C_5H_5^-$ 6e⁻		$C_5H_5^{\cdot}$ 5 e⁻
18e⁻		18 e⁻
Ionic counting method		Covalent counting convention

The electron count is of course identical to the alternative approach of distributing charges 'realistically' and treating metals as cations and ligands as anions. However, the neutral convention is simpler and more straightforward to use.

Ligands are classified according to the number of electrons they formally donate to the metal centre, following the covalent (electroneutral) counting convention:

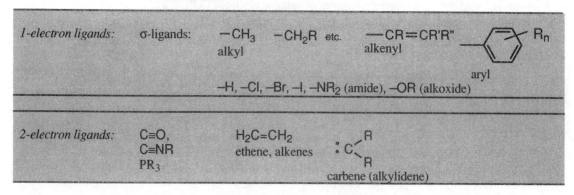

1-electron ligands: σ-ligands: —CH₃ —CH₂R etc. —CR=CR'R'' aryl Rₙ
alkyl alkenyl

—H, —Cl, —Br, —I, —NR₂ (amide), —OR (alkoxide)

2-electron ligands: C≡O, C≡NR PR₃   H₂C=CH₂ ethene, alkenes   :C⟨R R carbene (alkylidene)

3-electron ligands:

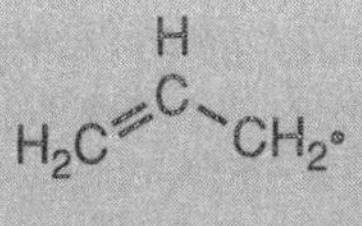

$$H_2C \overset{\overset{\textstyle H}{|}}{=} C \overset{\cdot}{=} CH_2$$

η^3-allyl,
η^3-enyl

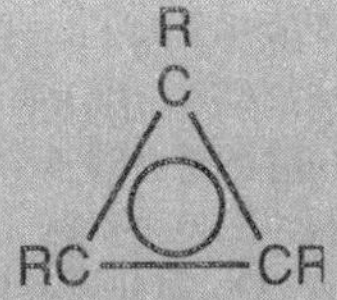

cyclopropenyl

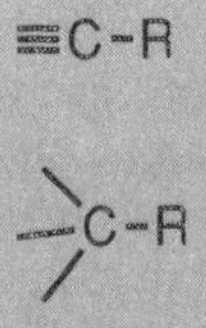

$\equiv C-R$

$\rangle C-R$

carbyne
(alkylidyne)

bridging halide

$N \equiv O$ nitric
oxide

4-electron ligands:

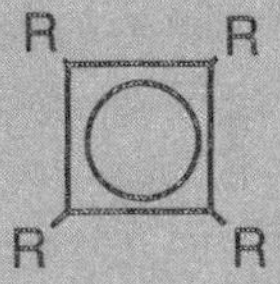

cyclobutadienyl

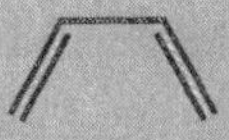

dienes
(conjugated)

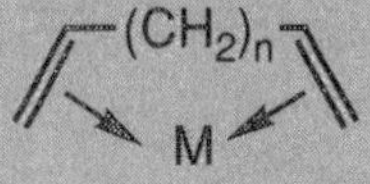

diolefins
(non-conjugated)

5-electron ligands:

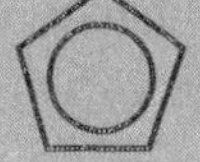

η^5-cyclopentadienyl

η^5-pentadienyl

6-electron ligands:

η^6-arene

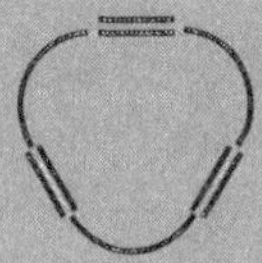

triene

7-electron ligands:

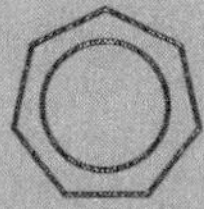

η^7-cycloheptatrienyl

8-electron ligands:

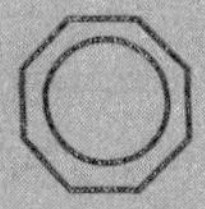

η^8-cyclooctatetraenyl

Note that ligands containing heteroatoms, such as Cl, OR, SR, NR$_2$, etc., can contribute more than one electron if lone electron pairs on the heteroatom are involved. A chloride ligand bridging two metals contributes three electrons (one σ-bond and one n-donor interaction), an –SR ligand bridging three metals is a 5-electron donor .

The vast majority of organometallic transition metal complexes obey the 18-electron rule. Notable exceptions are complexes of the late transition metals with d^8 configuration, such as Rh(I), Ir(I), Ni(II), Pd(II) and Pt(II): these have a strong preference to form square-planar 16-electron complexes. The reason is the increased stabilisation of the d-shell as the atomic number increases, so that e.g. the occupied d_{z^2} orbital no longer participates in ligand bonding, and stabilization through back-bonding (see Chapter 2) becomes less efficient. Characteristic types of d^8 complexes are for example:

With unsaturated ligands it is necessary to specify the number of carbon atoms which interact with the metal centre. The prefix η^n before the ligand formula implies bonding to n carbons, while μ_k indicates a ligand bridging k metal atoms. Individual numbering of ligand atoms may be required to describe more complicated structures:

$(\eta^4\text{-}C_7H_8)Fe(CO)_3$

$(\eta^3\text{-}C_7H_7)Fe(CO)_3$

$[\{(\eta^5\text{-}C_5Me_5)Re(CO)_2\}\{\mu_2\text{-}(\eta^2{:}\eta^2\text{-}C_6H_6)\}]$

$[(\eta^8\text{-}C_8H_8)Ti]_2[\mu\text{-}(1\text{-}4\eta{:}3\text{-}6\eta\text{-}C_8H_8)]$

$Fe_2(\mu_2\text{-}CO)_3(CO)_6$

The 18-electron rule is often not observed in cases of high steric hindrance, i.e. in complexes where the ligands are so bulky that not enough can be coordinated to allow the metal centre to achieve an electron count of 18. Examples are numerous. Complexes of this kind may be regarded as sterically saturated but electronically unsaturated. Nevertheless, they are frequently thermally quite stable—a measure of the importance of kinetic (versus thermodynamic) stabilization afforded by ligands with bulky substituents which provide an effective shield around the metal. Bulky ligands often allow the isolation of reactive species and complex types which may otherwise be inaccessible or highly unstable.

'Titanocene', $(C_5H_5)_2Ti$
14 VE; not detected

Decamethyltitanocene, $(C_5Me_5)_2Ti$,
an orange crystalline solid stable
at 0° C

unknown

isolable crystalline solid, 9 VE

2 Alkene complexes

In 1827 W. C. Zeise, a Danish pharmacist, refluxed K_2PtCl_4 in ethanol and obtained an unusual compound which, according to the elemental analysis, had the composition $KPtCl_3C_2H_4$. The true significance of this experiment was not realized until very much later: Zeise had made the first organometallic compound of a transition metal, an ethylene complex of platinum(II). The same complex is obtained under smoother conditions if K_2PtCl_4 is treated with ethylene in the presence of catalytic quantities of $SnCl_2$. In Zeise's original preparation the ethylene was formed by the dehydration of ethanol:

Zeise's choice of platinum was fortunate: as a rule olefin complexes of noble metals are air- and hydrolysis-stable, none more so than Pt, in contrast to complexes of first row and early transition metals. Alkene (olefin) complexes now exist of every transition metal and constitute one of the most important classes of coordination compounds. The development of the organometallic chemistry of olefins is closely connected with the rise of the petrochemical industry during the course of this century: olefins, in particular ethylene, are abundantly available from petrochemical feedstocks.

2.1 Synthesis

By ligand substitution

Oxidation with Me_3NO is often used to remove a CO ligand under mild conditions by oxidizing it to CO_2. If the reaction is carried out in MeCN, isolable highly reactive nitrile complexes result which are valuable synthetic intermediates. Cationic iron alkene complexes are synthetically useful since they are readily attacked by nucleophiles. The preparative method is also applicable to cationic alkyne, ketone and nitrile complexes.

Whereas alkenes as weak donor ligands are not usually able to displace phosphines to give alkene complexes (while the reverse reaction is usually facile), electron-rich early transition metals in low oxidation states favour alkenes as better π-acceptors:

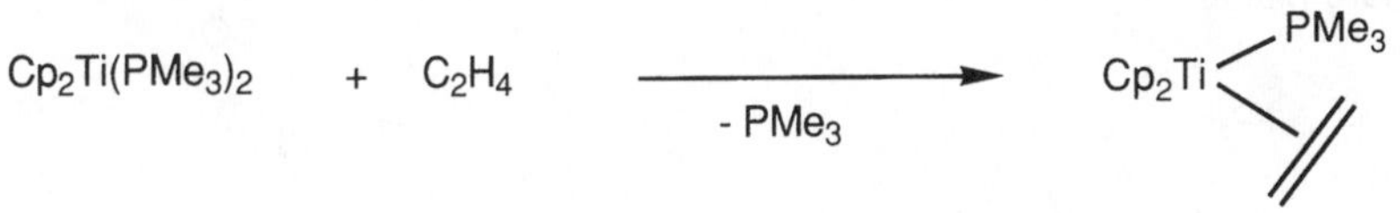

$$Cp_2Ti(PMe_3)_2 \quad + \quad C_2H_4 \quad \xrightarrow[-\text{PMe}_3]{} \quad Cp_2Ti$$

Cyclopentadienyl ligands may be substituted by alkenes if suitable anionic ligands are offered at the same time, for example by alkylation. For example, the 20 VE complex Cp_2Ni is converted into an 18 VE product in this way:

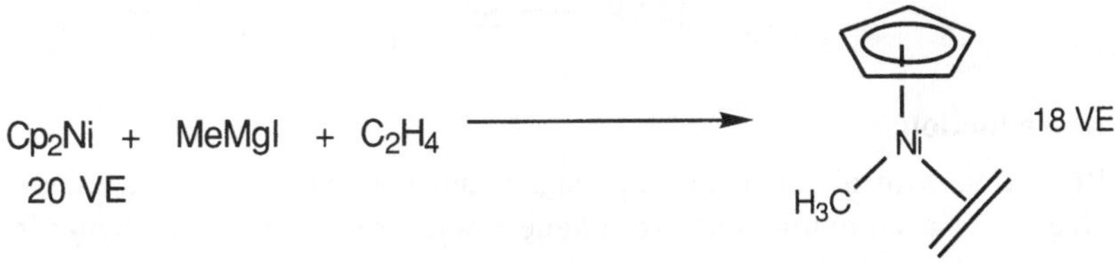

$$Cp_2Ni \; + \; MeMgI \; + \; C_2H_4 \quad \longrightarrow$$
20 VE

18 VE

Cyclododecatriene, a cyclotrimer of butadiene, forms a labile complex with Ni(0) which serves as a starting material for other nickel olefin complexes:

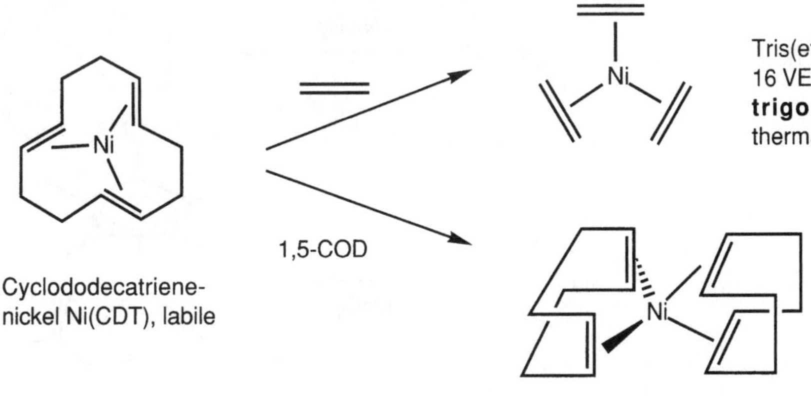

Cyclododecatriene-
nickel Ni(CDT), labile

1,5-COD

Tris(ethene) nickel
16 VE
trigonal-planar
thermally unstable

Bis(cyclooctadiene)nickel, Ni(COD)$_2$
stable yellow crystals,
18 VE, tetrahedral structure

By addition to coordinatively unsaturated complexes

$$IrCl(CO)(PPh_3)_2 \quad + \quad R_2C{=}CR_2 \quad \longrightarrow$$

16 VE, square-planar

18 VE

Photolysis of Cp_2ZrPh_2 leads to reductive elimination of biphenyl to give a highly reactive 14 VE intermediate, Cp_2Zr. For the thermolysis of Cp_2ZrPh_2 see p.41

$$Cp_2ZrPh_2 \xrightarrow[- \text{ Ph–Ph}]{h\nu} \left[Cp_2Zr \right] \longrightarrow$$

highly reactive
14 VE intermediate

By reduction

Reduction with alkali metals or magnesium reagents in the presence of alkenes is a common route to alkene complexes. Dienes are better π-acceptors than monoolefins and stabilize complexes of early transition metals in low oxidation states.

$$TiCl_4 + \text{dmpe} + \qquad \xrightarrow{\text{Na/Hg}}$$

$$Cp^*ZrCl_3 + \qquad \xrightarrow[- \text{ MgCl}_2]{}$$

$$Cp_2Co + C_2H_4 \xrightarrow[- \text{ KCp}]{K} \qquad \xrightarrow[- \text{ KCp}]{K} K^+ \left[Co(C_2H_4)_4 \right]^-$$

19 VE

18 VE; reactive complex with labile ethene ligands. Valuable catalyst.

18 VE, tetrahedral, extremely sensitive

Platinum metals are more readily reduced and generally form more stable organometallic complexes in low oxidation states. For example, refluxing ethanol is sufficient to reduce Rh(III) to Rh(I). The resulting olefin complexes have a structure similar to Zeise's salt (square-planar, d^8) and are useful starting materials.

$RhCl_3 \cdot n\,H_2O \quad + \quad C_2H_4 \quad \xrightarrow[\text{reflux}]{\text{EtOH}}$

16 VE, air-stable, square-planar, labile C_2H_4 ligands

By metal atom synthesis

Evaporation of the metal under high vacuum, and co-condensation of the metal vapour with potential ligands in excess on the walls of the evaporation vessel cooled to liquid nitrogen temperature often results in highly reactive compounds which may be inaccessible by other methods.

$Mo\,(g) \quad + \quad$

air-stable colourless crystals

$Fe\,(g) \quad + \quad \longrightarrow \quad Fe(1,5\text{-}COD)_2$

thermally unstable, 16 VE

$Ti\,(g) \quad + \quad$

$R = Bu^t$

2.2 Bonding of alkenes to transition metals

Olefins bind to transition metals via their π orbitals, by donating electron density into an empty metal d-orbital. However, since olefins are weak bases, the bond has to be stabilized by another bonding contribution: the donation of electron density from the metal to the olefin, more precisely into its π^* orbital which has the right symmetry for effective overlap with an occupied metal d-orbital. The metal therefore acts as both a Lewis acid (electron acceptor) and a Lewis base (electron donor) with respect to the olefin. This

For an explanation of the bonding of CO based on the Dewar–Chatt–Duncanson model see *Organometallics 1*, Chapter 2.

bonding concept was first proposed by Dewar and by Chatt and Duncanson (*'Dewar–Chatt–Duncanson model'*) and has proved its value in understanding metal–alkene bonding over the years.

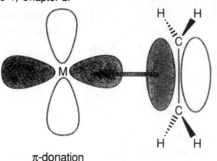

π-donation

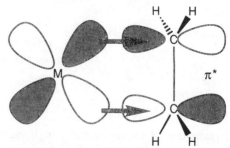

back–donation into olefin π^* orbitals

In analogy to CO, back-donation into C=C antibonding π^* levels leads to a weakening (lengthening) of the C=C bond and to a corresponding reduction in the C=C bond stretching frequency. However, unlike CO, this frequency lies in a less conveniently clear region of the spectrum; the bands are less intense and more difficult to identify and therefore of less diagnostic value.

C=C stretching frequencies (cm^{-1})		$rC{=}C$ (Å)
Free C$_2$H$_4$	1623	1.335
[(C$_2$H$_4$)$_2$Ag]BF$_4$	1584	
K[PtCl$_3$(C$_2$H$_4$)]	1516	1.375
CpRh(C$_2$H$_4$)$_2$	1493	

Back-bonding into olefin π^* levels has two structural consequences:

☞ lengthening of the C=C bond;
☞ reduction of the angles around C from *ca.* 120° (*sp^2* hybridized) towards angles typical of tetrahedral *sp^3*-C.

The extent to which back-bonding occurs depends on a number of factors:

☞ the energy of the occupied frontier orbitals of the ML$_n$ fragment to which the olefin binds;
☞ steric effects;
☞ electron acceptor strength of the olefin, i.e. the presence or absence of electron-withdrawing substituents.

These changes are most clearly revealed by determining the molecular structure of the complex by single-crystal X-ray diffraction.

The dependence of back-bonding on both the nature of the olefin and the metal is illustrated by olefin complexes of early transition metals. Metals from the left of the periodic table in low oxidation states, such as those of the titanium and vanadium triads, possess energetically high-lying occupied *d*-orbitals. Overlap with the π^* orbital of an olefin results in a significant energy gain, i.e. extensive back-bonding. In effect, excess electron density is offloaded onto the alkene. In the extreme, the C=C bond length reaches values more typical of C–C single bonds, with formation of two distinct metal–carbon σ-bonds, so that the resulting complex is more appropriately described as a **metallacyclopropane**. In effect, the coordination of an olefin has resulted in the oxidation of the metal:

π-complex   metallacyclopropane

Examples for such lengthening of the C–C bond of coordinated ethylene are Nb and Ta complexes, in the latter case the C–C bond has been extended to almost the value of a single bond, and the complex is best described as Ta(V).

By comparison, the ethylene ligand in Zeise's salt shows very little evidence of back-bonding and has a C=C bond length very close to that of free ethylene; the hydrogen atoms are almost in the plane of the C=C bond, and the C atoms remain sp^2 in character. Calculations have shown that the bond to Pt(II) relies predominantly on electron donation from ethylene to the metal, with only up to 25% of the total bond energy being contributed by back-bonding.

It should be noted that the ethylene ligand in Zeise's salt, as well as in other square-planar complexes of d^8 metals, adopts a conformation **perpendicular to the PtCl$_3$ plane**. Since there is no strong preference for the overlap of the ethylene π* orbital with the d_{xy} and the d_{xz} orbitals, steric interaction with the two *cis*-Cl ligands dictate the geometry and favour the perpendicular conformation.

Much more extensive back-bonding can be expected from zerovalent metals, such as the $Fe(CO)_4$ fragment. Its frontier orbitals give effective back-bonding stabilization with an in-plane olefin ligand. The resulting $(CO)_4M(alkene)$ complexes are trigonal bipyramidal, with the olefin in the trigonal plane for electronic reasons. $Fe(CO)_4$ is isolobal with CH_2 (see *Organometallics 1*, p. 31 ff.), and the ethylene complex may therefore be regarded as analogous to cyclopropane. In agreement with the metallacyclopropane description the C–C bond in $(CO)_4Fe(C_2H_4)$ is very long, 1.46 ± 0.06 Å.

Organometallic complexes are frequently fluxional; this is measured by variable-temperature NMR. Ethylene ligands for example are able to rotate with respect to the ethylene–metal vector. The size of these rotational barriers depends on the extent of back-bonding: little back-bonding as in $[PtCl_3(C_2H_4)]^-$ is reflected in a low rotational barrier (40–50 kJ mol^{-1}), while in complexes with extensive back-donation barriers of >100 kJ mol^{-1} have been found.

Even more extensive back-donation is found for the heavier transition metals. Since the stability of higher oxidation states increases within a triad with increasing atomic number, it is not surprising that the metallacycloalkane formulation of olefin complexes is even more appropriate here than with first row metals. In $Os(C_2H_4)(CO)_4$, and particularly in the dinuclear complex $Os_2(C_2H_4)(CO)_8$, the C–C bond lengths are typical for C–C single bonds, as in ethane. This is confirmed by the angles around the carbon atom.

This form of ethylene complexation is regarded as a model for the 'di-σ-bound' structure of chemisorbed ethylene on platinum metal surfaces.

Metallacyclopropane

Dimetallacyclobutane

Below is a comparison of distances and angles of free and coordinated ethylene:

	free C_2H_4	C_2H_6	$K[PtCl_3(C_2H_4)]$	$Os_2(C_2H_4)(CO)_8$
r_{C-C} [Å]	1.335(3)	1.532(2)	1.375(4)	1.5225(26)
H–C–H angle (°)	116.6	107.4	114.9	107.6
C–C–H angle (°)	121.7	111.5	121.1	111.9

Apart from the donor capacity of the ML_n fragment, the degree of back-bonding is dependent on the acceptor capacity of the olefins. This is strongly enhanced by electron-withdrawing substituents. F and CN substituents, in particular, render the alkene strongly electrophilic so that the resulting metal complexes generally adopt a metallacyclopropane structure, with almost tetrahedral carbon atoms:

In cases where structure determinations are not possible, NMR can provide information about the hybridization of carbon. The ^{13}C–^{13}C coupling constant is a good indicator of this: it changes from 67.6 Hz in free ethylene to 39 Hz in $Os(C_2H_4)(CO)_4$, in line with the reduction in bond order. Unfortunately, the more easily measured $^1J_{CH}$ is insensitive to these changes (c. 156 Hz).

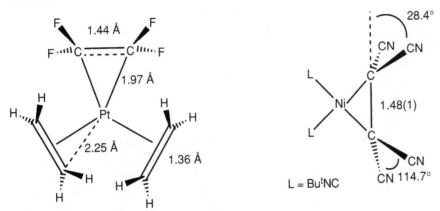

L = Bu^{t}NC

Nickel tetracyanoethylene complex: Ni(0) or Ni(II)?

Note that the substituents on carbon are no longer in the olefinic plane and give small X–C-X angles of close to 110°. The angular deformation is a more sensitive indication of back-bonding than the elongation of the C=C bond.

A comparison between C_2H_4 and C_2F_4 within the platinum complex $Pt(C_2H_4)_2(C_2F_4)$ confirms the bonding concept: short Pt–C and long C–C distances indicate tight bonding to C_2F_4, and little back-bonding to the ethylene ligands.

These alkene complexes of (formally) zerovalent nickel and platinum (d^{10}), as well as a host of similar compounds $L_2M(alkene)$ (L = phosphine or phosphite, M = Ni, Pd, or Pt) adopt a conformation which is at first surprising: the alkene ligands are coordinated 'in-plane'. By contrast, it was shown above that the complexes of d^8 ions such as Rh(I) and Pt(II) contain the alkene bonded perpendicular to the coordination plane, to minimize steric interactions.

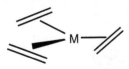

stable conformation,
maximum back–donation

not found with d^{10} metals

The reason is electronic: the d_{xz} and d_{yz} orbitals in an angular ML_2 fragment are of very different energy, and the energetically high-lying occupied d_{xz} orbital leads to more efficient back-bonding. Consequently, the alkene ligand has to adopt an 'in-plane' conformation to maximise the overlap with d_{xz}.

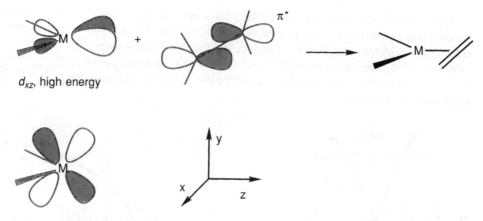

d$_{xz}$, high energy

d$_{yz}$, lower energy orbital

A frequently used olefinic ligand is 1,5-cyclooctadiene (COD). In principle it behaves like two ethylene molecules tied together. It is therefore able to act as a chelating ligand, and it is evident that cyclooctadiene complexes are more stable than ethylene complexes as a result (**chelate effect**). Nevertheless, COD complexes show a very similar reactivity to ethylene complexes and the COD ligand is quite easily displaced, although, as the comparison of trigonal-planar $Ni(C_2H_4)_3$ with the tetrahedral $Ni(COD)_2$ shows (p. 7), it sometimes imposes certain geometric preferences.

A special case of olefin complexes are the adducts of buckminsterfullerene, C_{60}. Fullerenes are football-shaped carbon networks of a variable number of 6-membered, and twelve 5-membered rings, as required by the spherical geometry. Unlike graphite, the 6-membered rings have little aromatic character. The compound behaves like a polyolefin with significant π-acceptor capacity and forms complexes e.g. with PtL_2 fragments (L = PEt_3 or PPh_3). The metal coordinates to the most electron-rich C–C bond linking two 6-rings, with considerable elongation of the C–C distance. One C_{60} can bind up to six PtL_2 units. Similar complexes are obtained from a higher homologue, C_{70}.

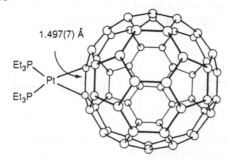

1.497(7) Å

$(Et_3P)_2Pt(C_{60})$

Bonding in diene complexes

Compared to monoolefins, *dienes* have energetically lower lying π^* orbitals and are consequently **more powerful π-acceptors**. This is evident from the examples presented on p. 8ff, which show the synthesis e.g. of diene complexes of titanium(0)—a metal that really prefers the oxidation states III and IV—for which no monoolefin analogues are known.

On coordination to a metal dienes usually adopt the s-*cis*-conformation, as in the classical representative of this class, (butadiene)iron tricarbonyl. This compound was one of the very early examples of organometallic π-complexes, first made in 1938 by O. Roelen at BASF. The main structural feature of this complex, apart from its 'half-sandwich' structure (see Chapter 4), is the equality of the C–C bonds in the coordinated butadiene:

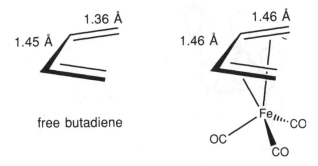

This type of diene bonding may be described by resonance forms:

In molecular orbital terms, structure **B** is equivalent to the population of the LUMO (Ψ_3) of butadiene by metal *d*-electrons:

	Overlap with metal orbital
Ψ_4	d_{xy}
Ψ_3 LUMO C2-C3 bond strengthened by back-donation	d_{xz}, p_x
Ψ_2 HOMO C2-C3 bond weakened	d_{yz}, p_y
Ψ_1	d_{z^2}, p_z, s

Early transition metals have a strong preference for the highest possible oxidation state and form complexes in which back-bonding into Ψ_3 is extensive; consequently there is a tendency to adopt the metallacyclopentene structure **B**. Typical indicators for this coordination mode are short bonding distances between metal and the terminal carbons, a short internal C=C bond and elongated terminal C–C bonds, which contrasts with the even bond length distribution of the diene ligand in $(C_4H_6)Fe(CO)_3$.

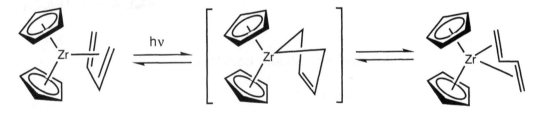

2.30 Å
CH_2
CH_2
1.45 Å
1.40 Å
Me
2.59 Å
Me

Metallacyclopentene structure
of Cp$_2$Zr(dimethylbutadiene)

While this dimethylbutadiene complex is stable enough to be crystallographically characterized, the behaviour of the parent complex Cp$_2$Zr(butadiene) is more interesting: the complex is an unusual example for a photochemically driven equilibrium between the s-*cis* and s-*trans* diene conformations. The reaction proceeds via a metallacyclopentene intermediate which, in this case, is not the thermodynamic minimum.

hv

There are only few cases of such s-*trans*-coordinated dienes. In at least one example, CpMo(NO)(butadiene), the s-*trans* configuration is actually preferred. This appears to be the exception, and as a general rule the rearrangement to the thermodynamically favoured s-*cis* isomer is facile. However, in cases where *cis/trans* isomerization is slow it has been possible to prepare each isomer stereoselectively:

The bridging of two metal centres has also been observed in di- and trinuclear complexes, as in $Mn_2(CO)_8(C_4H_6)$ and $Os_3(CO)_{10}(C_4H_6)$.

2.3 Reactivity of alkene complexes

Ligand substitution

As relatively weak σ-donors and π-acceptors, alkenes are easily displaced, for example by phosphines. The volatility of the ligand makes ethylene complexes particularly labile; they are usually best stored under an ethylene atmosphere. Complexes of 1,5-cyclooctadiene are more stable, without loss of reactivity, and are commonly used as synthetic reagents and catalyst precursors.

Facile ligand dissociation means that olefin complexes can sometimes be regarded as the storage form of 'naked metal' which is highly reactive, e.g. towards oxidative addition:

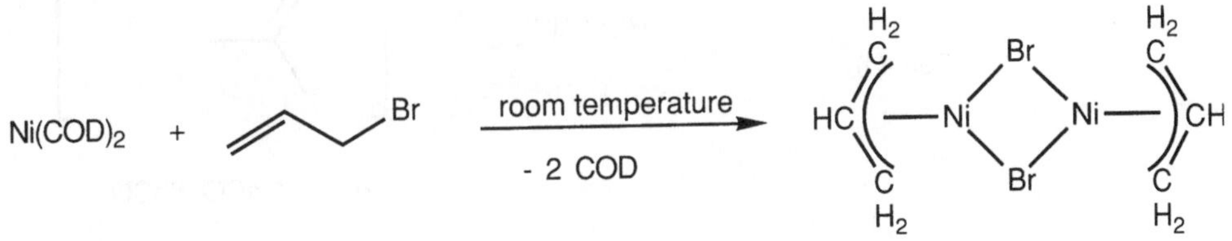

Another synthetically valuable olefin complex is $Pd_2(dba)_3$. Dibenzylideneacetone (dba, 1,5-diphenyl–1,4-pentadien-3-one) is a diene with particularly good π-acceptor ability and gives air-stable complexes in cases where COD complexes are not obtainable. $Pd_2(dba)_3$ is a ready source of palladium(0):

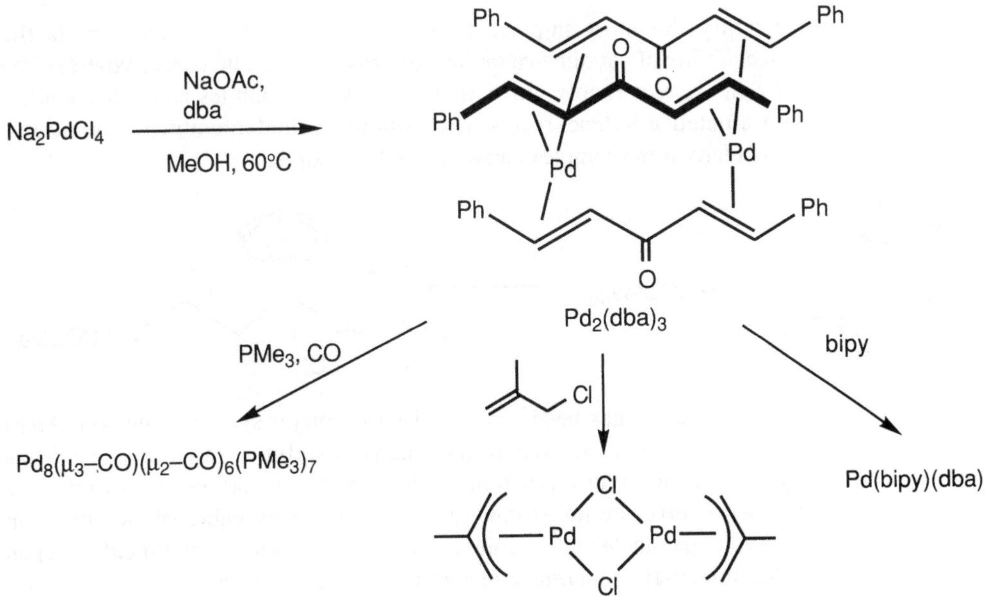

Reactions with electrophiles

The reaction of alkene complexes with electrophiles, for example HX, generally leads to the reductive loss of the alkene and the formation of the corresponding L_nMX_y complex. In favourable cases, however, the intermediate alkyl can be isolated. Protonation of allene and diene complexes gives isolable allyl products.

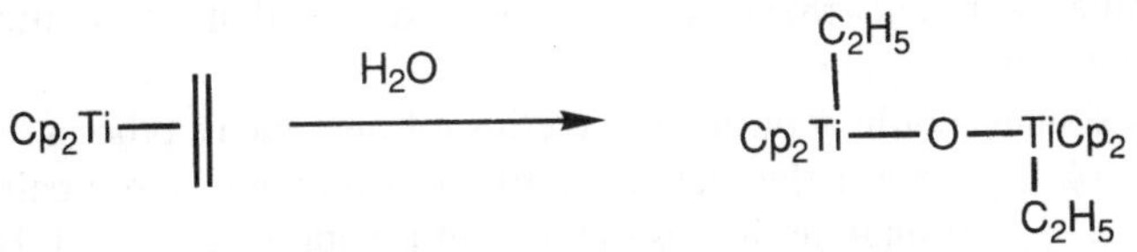

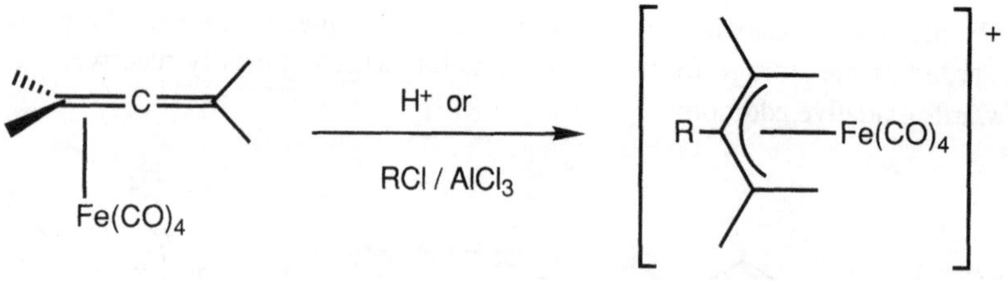

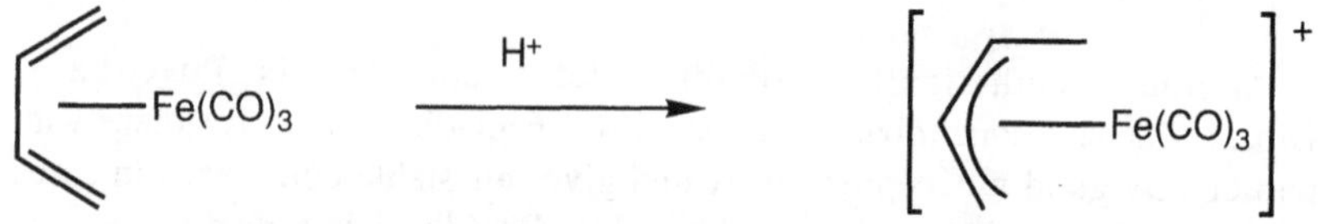

R = H, MeCO, PhCO

Reactions with nucleophiles

Certainly the most important aspect of metal-olefin chemistry is the susceptibility of coordinated olefins towards nucleophilic attack. Whereas free ethylene is remarkably unreactive towards nucleophiles and electrophiles, coordinated ethylene reacts very smoothly under ambient conditions, particularly if the complex carries a positive charge:

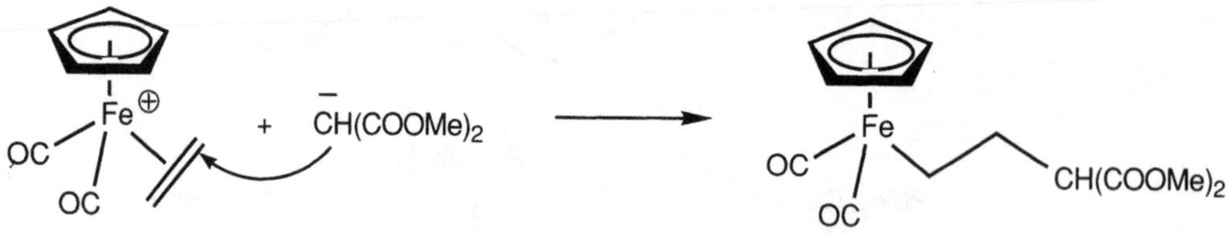

This reaction has been developed into a major synthetic method. Apart from stoichiometric reactions, many catalytic reactions, such as hydrogenation, hydrocyanation or the Wacker oxidation of ethylene (see below) involve the transformation of the olefin by nucleophilic attack. In principle, the nucleophile can react with the coordinated olefin either via an *intra*molecular or an *inter*molecular (external) pathway:

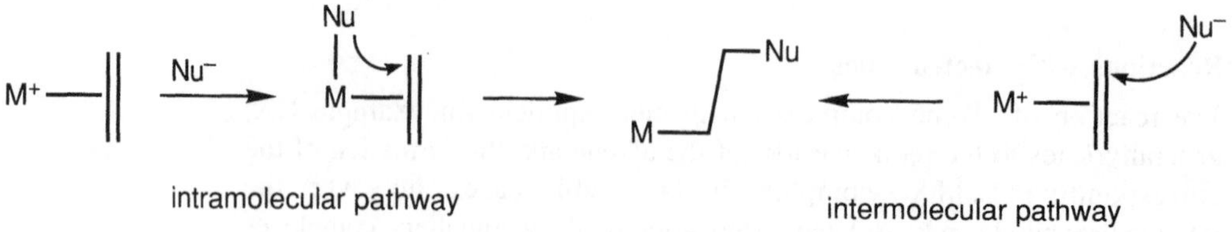

intramolecular pathway

intermolecular pathway

Which one of these options prevails in a given case may be difficult to determine, although there is growing evidence that most attacks are *intramolecular*.

Why should coordination activate olefins towards such a process? After all, a nucleophilic attack is the interaction of an incoming electron pair with an empty receptor orbital, in the case of an olefin with its π^* orbital. However, in the complex this π^* orbital is already partly populated by back-bonding from the metal, a process which ought to result in a reduced ability to react with nucleophiles. The opposite is the case.

The answer lies in the fact that molecules and bonds are not rigid but oscillate around their equilibrium positions. If an ethylene ligand shifts sideways with respect to the metal, the C–C bond becomes polarized. In an extreme case the carbon atom at one end could be thought to assume cationic character:

It is obvious that such a polarization enhances the susceptibility towards a nucleophilic attack. The lack of reactivity of free ethylene is due to a lack of polarity within the molecule.

Nucleophilic olefin reactions in catalysis 1. Olefin hydrogenation

One of the simplest and most fundamental catalytic reactions of olefins is hydrogenation. It is catalysed by a multitude of homogeneous and heterogeneous catalysts, and the mechanism has been thoroughly investigated. In all cases, the essential step is the 'activation' of dihydrogen by the transition metal (oxidative addition), followed by the transfer of a hydride ligand to a coordinated alkene. The process is illustrated for the case of one of the best-known catalysts, $RhCl(PPh_3)_3$ *(Wilkinson's catalyst)*.

It is essential for the establishment of a catalytic cycle to generate coordinatively unsaturated intermediates which are able to bind hydrogen and the alkene. In this case the triphenylphosphine ligand is labile and dissociates readily in solution to leave intermediates with vacant coordination sites. The reactive ligands, in this case the hydride and the olefin or alkyl, have to occupy mutually *cis* positions; this principle has been discussed in detail before (*Organometallics 1*, p. 60 ff.).

Unlike heterogeneous catalysts, homogeneous rhodium and ruthenium complexes are highly chemo- and regioselective.

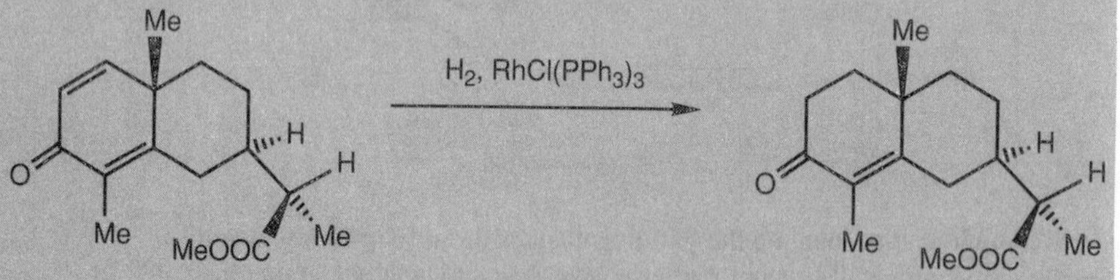

Chiral phosphine ligands allow the asymmetric hydrogenation of prochiral alkenes with high enantiomeric excess; examples of these are:

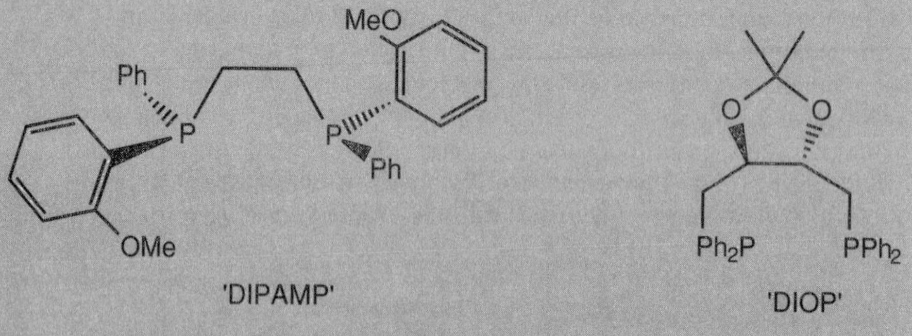

Enantiomeric excess (e.e.) is the excess concentration of one enantiomer over the other, so an e.e. of 95% implies an enantiomer ratio of 97.5:2.5.

This method is frequently used in the laboratory and commercially for the synthesis of L-dihydroxyphenylalanine (L-dopa), an aminoacid for the treatment of Parkinson's disease. This process uses a phosphine with a chiral P atom ('DIPAMP').

Nucleophilic olefin reactions in catalysis 2. Hydrosilation

The addition of silanes R_3SiH to C=C double bonds ('hydrosilation') is catalysed by a number of noble metals, such as Wilkinson's catalyst [$RhCl(PPh_3)_3$], cobalt and iridium.

Platinum complexes are particularly active (unusually for a third row transition metal). H_2PtCl_6 (*Speier's catalyst*) is frequently used. It has been shown that the Pt(IV) precursor is reduced to Pt(0) alkene complexes under catalytic conditions; these are able to oxidatively add silanes to give $L_2Pt(H)(SiR_3)$ which reacts with alkenes under insertion in a manner analogous to alkene hydrogenation.

$H_2Pt^{IV}Cl_6$ precursor → vinylsilane → isolable Pt^0 intermediate → R_3SiH →

Nucleophilic olefin reactions in catalysis 3. The Wacker process

The *Wacker process* involves the oxidation of ethylene to acetaldehyde and is based on the combination of three well-known reactions;

1. the oxidation of ethylene by aqueous Pd^{2+} (known since 1894);
2. the copper catalysed oxidation of Pd^0 to Pd^{2+};
3. the air oxidation of Cu^+ to Cu^{2+}.

These three reactions combine to hydrate and oxidize ethylene to acetaldehyde and to regenerate the palladium(II) catalyst.

Acetaldehyde is a valuable industrial intermediate, mainly for the synthesis of acetic acid. The process has been operated by Wacker Chemie (a Hoechst subsidiary) since the late 1950s, though it is now being superseded by the methanol-based Monsanto process (*Organometallics 1*, p. 66).

(1) $C_2H_4 + PdCl_2 + H_2O \longrightarrow CH_3CHO + Pd^0 + 2\,HCl$

(2) $Pd^0 + 2\,CuCl_2 \longrightarrow PdCl_2 + 2\,CuCl$

(3) $2\,CuCl + 2\,HCl + 1/2\,O_2 \longrightarrow 2\,CuCl_2 + H_2O$

$C_2H_4 + 1/2\,O_2 \longrightarrow CH_3CHO$

A key step is the nucleophilic attack of water or OH^- on coordinated ethylene. There has been much debate about this mechanism, in particular whether this attack by water is external or intramolecular. Recent elegant stereochemical studies support an *external* nucleophilic attack on coordinated ethylene by water, rather than the often postulated intramolecular migration of OH^-. The scheme below gives the current idea about the mechanism.

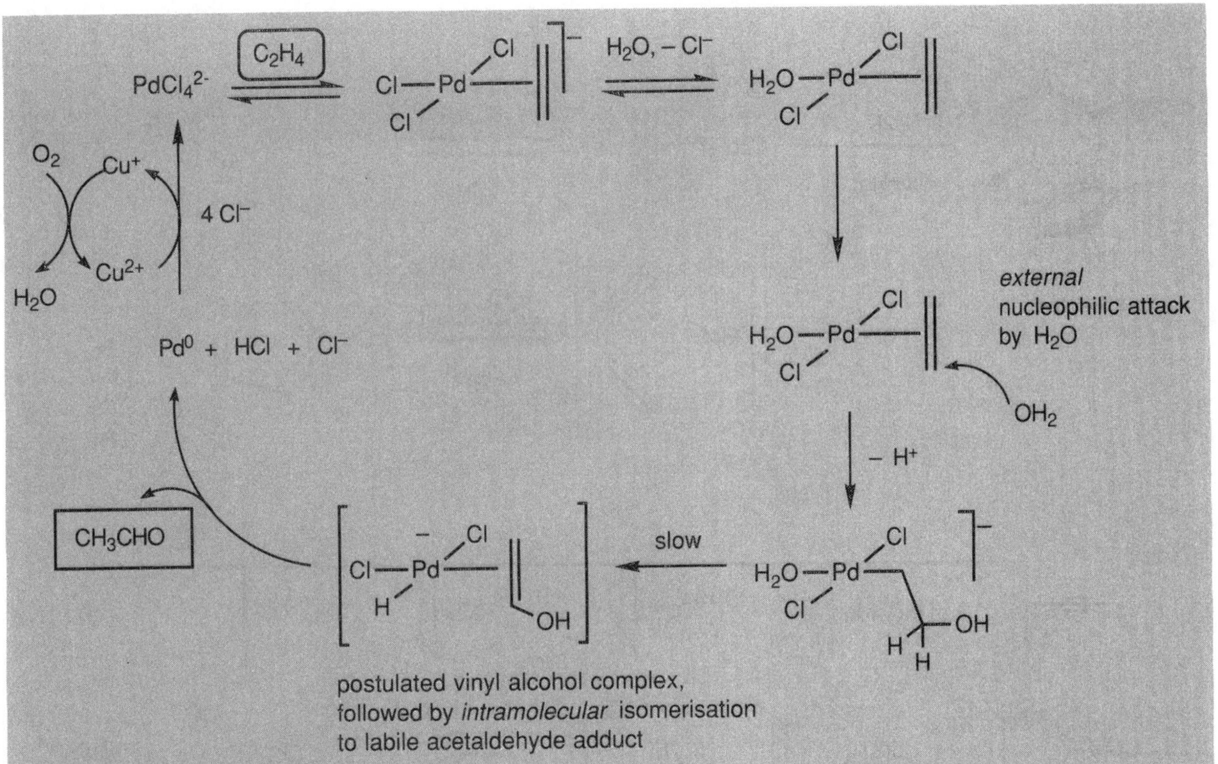

$PdCl_4^{2-}$

C_2H_4

$Cl-Pd\overset{Cl}{\underset{Cl}{\diagup}}\|$

$H_2O, -Cl^-$

$H_2O-Pd\overset{Cl}{\underset{Cl}{\diagup}}\|$

O_2 Cu^+

Cu^{2+} $4\,Cl^-$

H_2O

$Pd^0 + HCl + Cl^-$

$H_2O-Pd\overset{Cl}{\underset{Cl}{\diagup}}\|$

*external
nucleophilic attack
by H_2O*

OH_2

$-H^+$

CH_3CHO

$\left[Cl-Pd\overset{Cl}{\underset{H}{\diagup}}\overset{\|}{\underset{OH}{}}\right]^-$

slow

$\left[H_2O-Pd\overset{Cl}{\underset{Cl}{\diagup}}\overset{}{\underset{H}{\underset{H}{\diagdown}}}OH\right]^-$

postulated vinyl alcohol complex,
followed by *intramolecular* isomerisation
to labile acetaldehyde adduct

The Pd/Cu catalysed oxidation of olefins can be applied generally to many substrates. If acetate is used as the nucleophile instead of water, alkenyl acetates result which may be generated stereoselectively, for example:

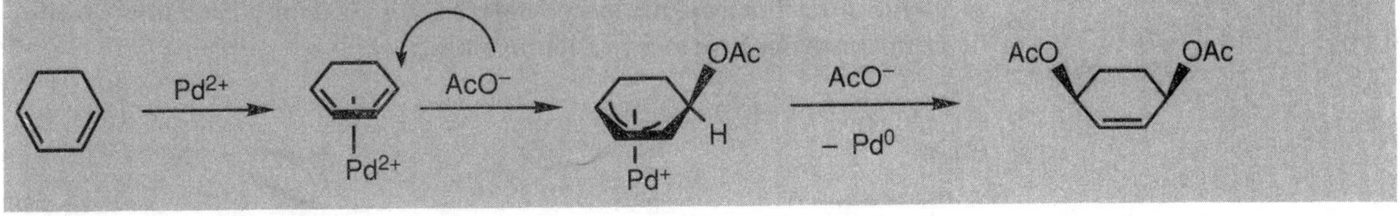

Pd^{2+} — Pd^{2+} — AcO^- — Pd^+, OAc, H — AcO^-, $-Pd^0$ — AcO, OAc

Metallacycles and C–H activation of alkenes

As discussed above, early transition metals have a propensity for extensive back-bonding and give alkene complexes which are often best described as metallacyclopropanes. These are often highly reactive and can undergo ring-expansion with a second alkene molecule to give a metallacyclopentane. The reaction, illustrated below for $Cp*_2Ti(C_2H_4)$ ($Cp* = \eta^5-C_5Me_5$), is **reversible**: in the absence of an excess of ethylene, for example, the mono-olefin complex is regenerated. In a similar way, the metallacyclopentane reacts with alkynes under loss of one ethylene to give the more stable metallacyclopentenes. With heteroalkenes and heteroalkynes (aldehydes, nitriles, CO_2) a range of cyclic products is obtained which exemplify the synthetic potential of these compounds. Exposure of $Cp*_2Ti(C_2H_4)$ to hydrogen gives $Cp*_2TiH_2$ under very smooth conditions; the by-product is ethane.

An interesting and important reaction is the activation of C–H bonds of alkenes. Such an activation (C–H bond breaking) is particularly facile with allylic C–H bonds. Propene complexes, for example, are often readily deprotonated by base to give allyl complexes.

If the energy difference between the olefin complex and the C–H activation product is small, an equilibrium between both species may be observed:

See also: diene oligomerizations involving allyl complexes (Chapter 4).

Palladium olefin complexes are particularly readily converted into allyls on treatment with base (see Chapter 5); this is a widely applicable standard reaction for the conversion of palladium alkene complexes into allyl compounds and much used in organic synthesis.

The *olefinic* C–H bonds of ethylene and other 1-alkenes, too, are not as inert as was thought for a long time. This activation results in vinyl complexes. Two types of metal systems are known to be able to achieve this reaction: highly electrophilic compounds of scandium or the lanthanides, and certain complexes of iridium.

The exchange of alkyl ligands by σ-bonded alkenyls or aryls is known as 'σ-bond metathesis' (see *Organometallics 1*, p. 75).

$M = Sc, Lu$

The formation of an iridium vinyl complex is particularly surprising since iridium is well known for forming very stable alkene complexes. The iridium vinyl hydride complex is not an intermediate in the formation of the iridium ethylene complex which is formed at the same time since the reductive elimination to give the olefin compound only proceeds under forcing conditions.

Isomerization of olefins

A C=C double bond can be shifted along the backbone of an olefin to give mixtures of isomeric terminal and *cis*- and *trans*-internal olefins. Two types of catalysts can be distinguished: metal hydides which react with alkenes to give alkyls (*hydride mechanism*), and low-valent metal complexes which form metal allyl species (*allyl mechanism*).

Allyl mechanism:

Hydride mechanism:

Since isomerizations are equilibrium processes, the thermodynamically most stable isomer is usually formed predominantly, although during the initial stages of the reaction the kinetic product can often be isolated (e.g. the *cis*-alkene instead of the *trans*).

The mechanism of olefin isomerization resembles that of olefin hydrogenation, and indeed isomerization is often found as a side reaction in slow hydrogenations.

Under certain circumstances isomerizations can give products which are apparently derived from the *thermodynamically least stable* olefin isomer. An example is the **hydrozirconation** of olefins: in the resulting zirconium alkyl the metal is always attached to the terminal carbon. The reason for this course of isomerization is the relative instability of secondary and tertiary metal–alkyl bonds compared to the sterically less encumbered Zr–n-alkyl bond. The reaction has considerable synthetic potential: the resulting products correspond to a functionalization of the alkene in the (thermodynamically disfavoured) *anti-Markovnikov* sense, i.e. they are complementary to classical methods of alkene functionalizations.

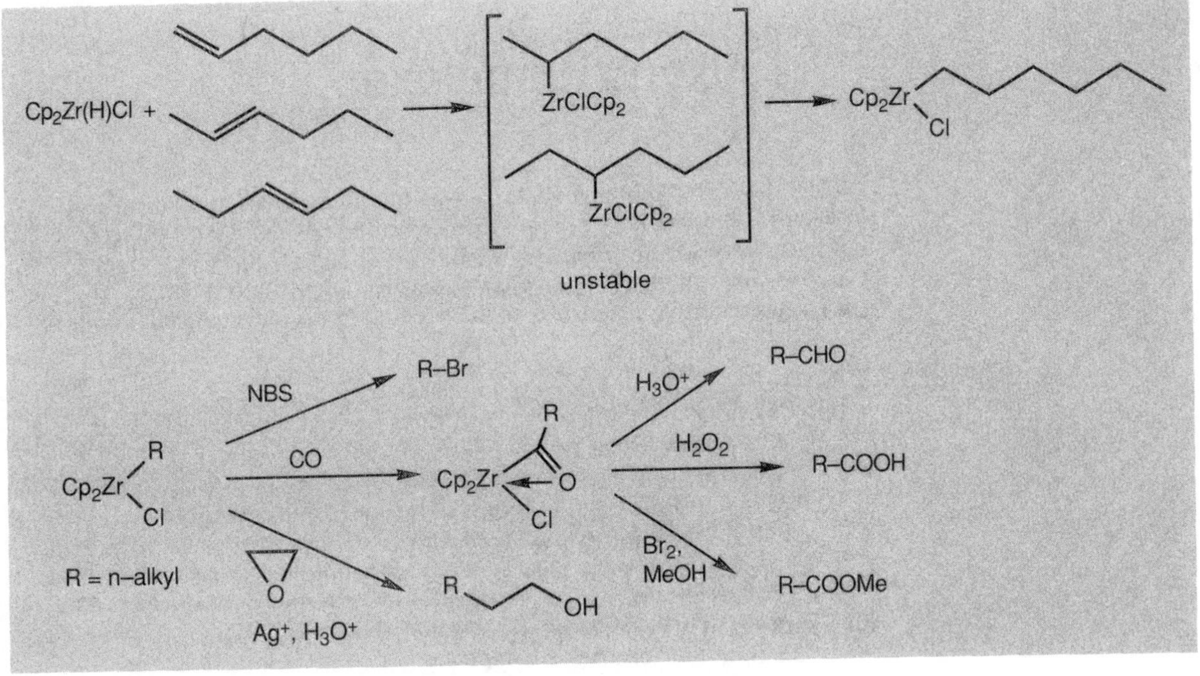
Cp2Zr(H)Cl +
ZrClCp2
ZrClCp2
unstable
Cp2Zr
Cl
NBS
R–Br
CO
R
Cp2Zr
Cl
R = n–alkyl
Ag+, H3O+
R
OH
Cp2Zr
O
Cl
H3O+
R–CHO
H2O2
R–COOH
Br2,
MeOH
R–COOMe

3 Alkyne complexes

Alkynes (acetylenes) form complexes with transition metals in a similar way to alkenes, and similar bonding schemes can be applied. C≡C bonds are, however, very energy rich and usually more reactive than C=C bonds. Characteristic differences in the bonding of alkynes to metals compared to alkenes are:

☞ alkynes are stronger π-acceptors than alkenes;
☞ alkynes have two orthogonal π-systems and can act as 2- as well as 4-electron donor ligands;
☞ alkynes frequently form integral parts of metal clusters, with loss of the high-energy C≡C triple bond;
☞ alkynes frequently undergo insertion reactions and are readily cyclotrimerized to give arenes.

3.1 Synthesis of alkyne complexes

Synthetic methods resemble those of alkene compounds; ligand substitutions with and without reducing conditions prevail. Some examples:

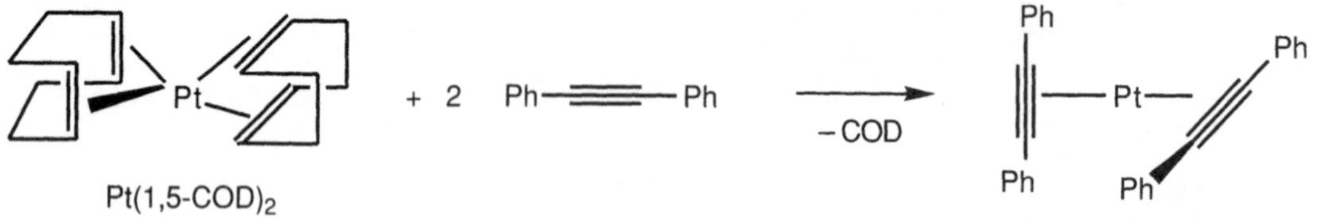

Pt(1,5-COD)$_2$ + 2   Ph———Ph $\xrightarrow{\text{− COD}}$

orthogonal alkyne ligands:
2- or 4-electron donors?

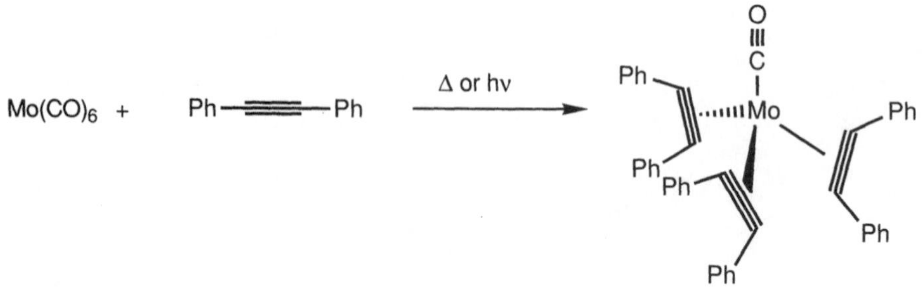

Mo(CO)$_6$ +   Ph———Ph $\xrightarrow{\Delta \text{ or } h\nu}$

14, 16 or 18 VE complex?

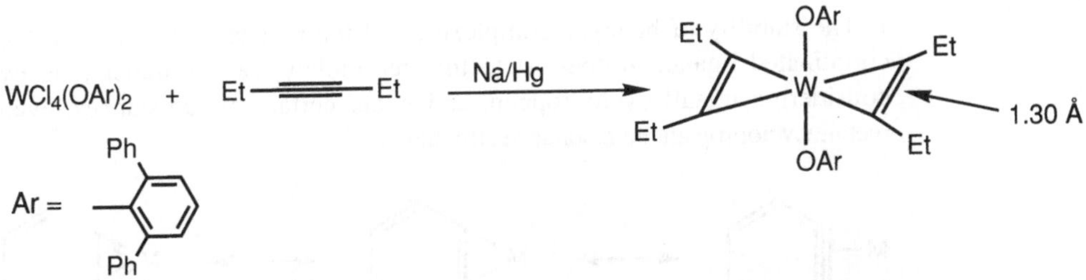
WCl4(OAr)2
+
Et—≡—Et
Na/Hg
OAr
Et
Et
W
Et
Et
OAr
1.30 Å
Ar =
Ph
Ph

The reductive coupling of two CO ligands provides an unusual and instructive example of the formation of an alkyne within the complex coordination sphere. Such reductive C–C coupling reactions, possibly followed by hydrogenation of the resulting alkyne ligand, are being discussed for the conversion of synthesis gas (CO/H_2) to hydrocarbons (*Fischer–Tropsch reaction,* see *Organometallics 1*, p. 33).

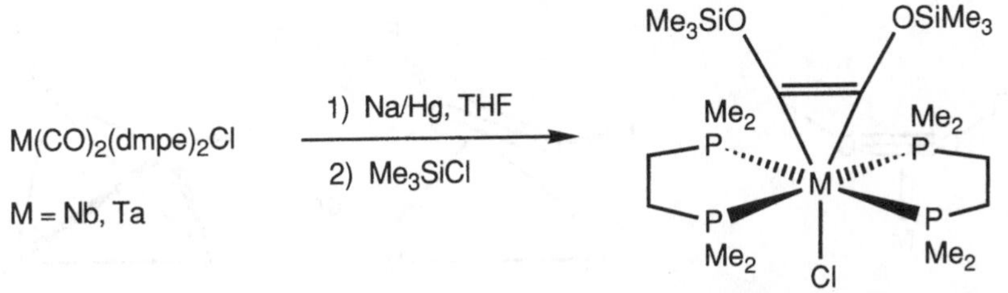
M(CO)2(dmpe)2Cl
M = Nb, Ta
1) Na/Hg, THF
2) Me3SiCl
Me3SiO
OSiMe3
Me2
Me2
P
P
M
P
P
Me2
Me2
Cl

A special case of an alkyne is benzyne, C_6H_4. The free molecule would be highly strained and is not stable; however, the benzyne ligand can be generated within the coordination sphere of a metal. The ring strain makes benzyne complexes highly reactive and synthetically useful for coupling reactions.

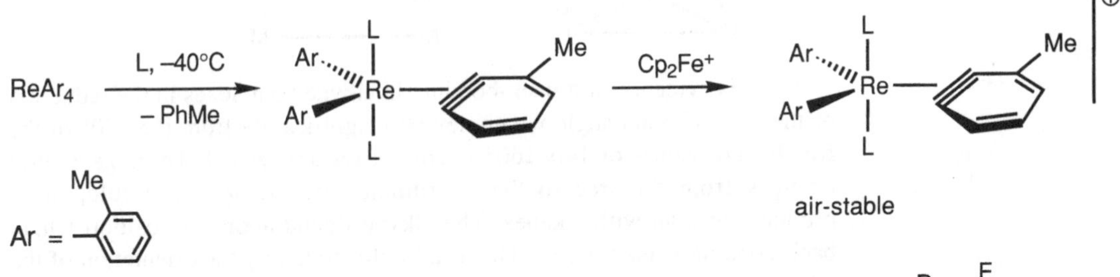
ReAr4
L, –40°C
– PhMe
L
Ar
Ar
Re
L
Me
Cp2Fe+
L
Ar
Ar
Re
L
Me
⊕
air-stable
Ar =
Me

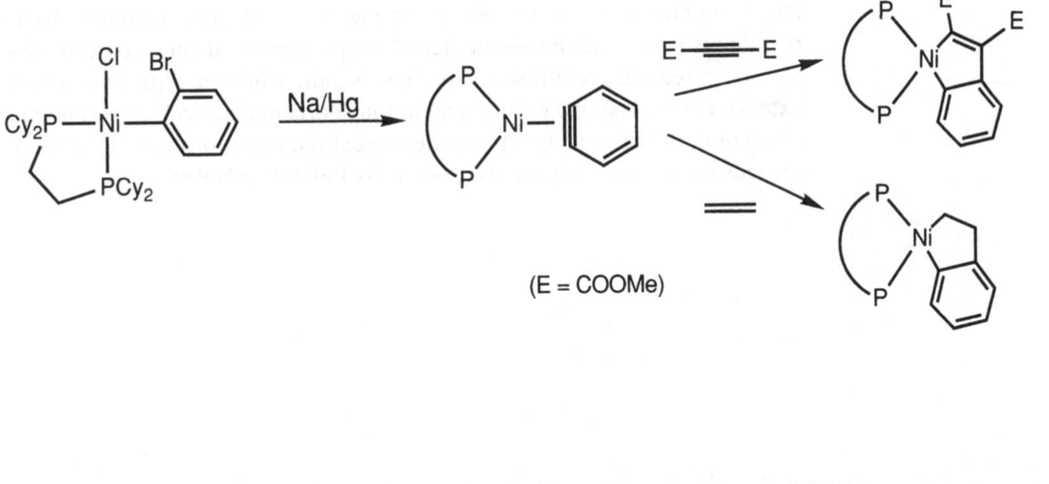
Cl
Br
Cy2P
Ni
PCy2
Na/Hg
P
Ni
P
E—≡—E
P
E
E
Ni
P
P
Ni
P
(E = COOMe)

The stability of benzyne complexes, and the reduced ring strain in the coordinated ligand, in contrast to the free moiety, can be understood by considering metallacyclopropene, and under certain circumstances, even metallacyclopropadiene resonance structures:

3.2 Bonding in alkyne complexes

Alkyne ligands can adopt a variety of coordination modes in mono- and polynuclear complexes, such as:

There is evidence of π back-bonding in alkyne complexes in the reduction of the R–C–C bond angle which deviates significantly from the 180° of the free ligand; values of 140–160° in complexes are typical. These geometric changes from the free to the coordinated ligand are, as a rule, more pronounced than with alkenes. The alkyne ligand adopts a position where back-bonding is maximized. This can be illustrated by the orientation of the alkyne ligand in Pt(II) and Pt(0) complexes: in the first instance, back-bonding is weak, and the alkyne ligand adopts a perpendicular conformation for steric reasons, reminiscent of Zeise's salt, while in Pt(0) compounds extensive back-bonding enforces an in-plane orientation, with corresponding elongation of the C≡C bond and the typical reduction of the C–C–R angle towards values more characteristic of sp^2-hydridized carbon atom.

Bu^t / 163°

Cl C

(p-tol)NH$_2$ — Pt — ‖ ← 1.24 Å

Cl C

Bu^t

perpendicular coordination to Pt^{II}:
C≡C short, wide C-C-R angle

Ph$_3$P. ...··Ph

C

Pt 140°

Ph$_3$P C ← 1.32 Å

Ph

in-plane bonding to Pt^0:
C≡C elongated, smaller C-C-R angle

In dinuclear complexes such as $Co_2(CO)_6(C_2R_2)$ the C–C bond of the alkyne ligand is even more elongated. The hydrocarbon ligand is very tightly bonded, and the compound is more adequately described as a tetrahedral cluster with a Co_2C_2 core, rather than an alkyne adduct.

1.35 - 1.37 Å

R

R C

C

$(OC)_3Co$ ——— $Co(CO)_3$

$Co_2(CO)_8$ + R–C≡C–R ⟶

ca. 1.20 Å

The structure can be rationalized by considering the cluster as composed of two $Co(CO)_3$ fragments and two carbyne moieties, R–C, rather than an alkyne ligand. Such a carbyne unit is *isolobal* with $Co(CO)_3$, that is both moieties have frontier orbitals of similar symmetry (see *Organometallics 1*, p. 32); it is also isoelectronic, neglecting core electrons, since the number of frontier electrons is three in either case.

Reduction of C≡C bond order in tetranuclear Co clusters leads to a reduction in the C≡C stretching frequency in the IR spectrum: for R = Me from 2313 to 1633 cm^{-1}.

R–C≡C–R ⟶ R–C ⟷ OC–Co–OC / OC

Cobalt complexes of this kind can be used in synthetic schemes for protecting triple bonds. The alkyne ligand can be set free by oxidative removal of the metal, e.g. by treatment with iodine.

3.3 Reactivity of alkynes

Alkynes, like alkenes, insert readily into metal–hydride bonds to give vinyl complexes:

L
|
Cl—Pt—H   R≡R ⟶
|
L

H R
\ /
C
Cl—Pt···L
/ \
L R

Acetylene precedes ethylene as the major chemical feedstock, and many of its transformations involve organometallic species. Much of the early acetylene and olefin chemistry involving organometallic catalysts was developed by W. Reppe (BASF) in the 1940's.

Vinyl compounds are also formed on protonation of alkyne complexes. Alkyne compounds of electron-rich metals such as rhodium have a very varied chemistry, and with alkynes containing activated C–H bonds, rearrangements to allene and vinylidene complexes have been observed:

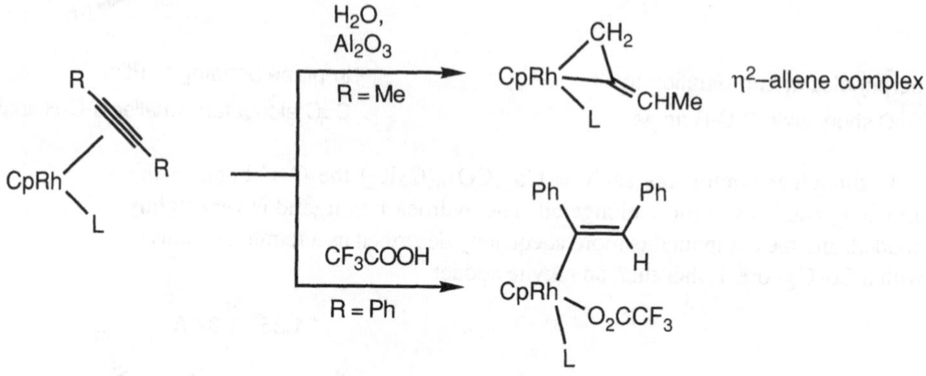

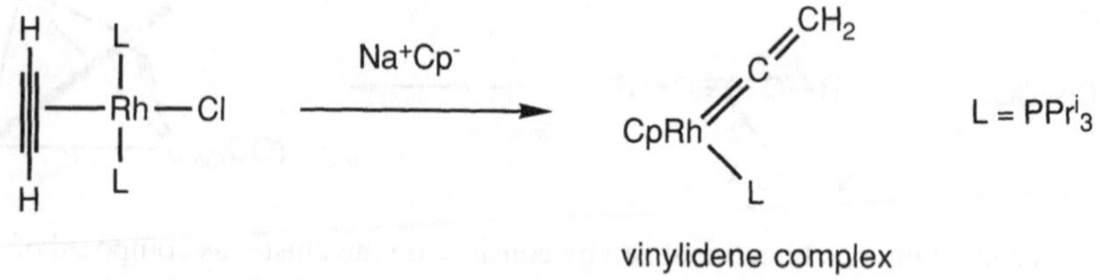

Alkynes undergo numerous insertion reactions into M–C and M–H bonds, particularly if the alkyne is electron-withdrawing [e.g. $C_2(COOMe)_2$ or $C_2(CF_3)_2$]. The insertion of alkynes into Pd–Cl bonds is surprisingly facile and reversible; the reaction models the first steps of the catalytic alkyne oligomerization:

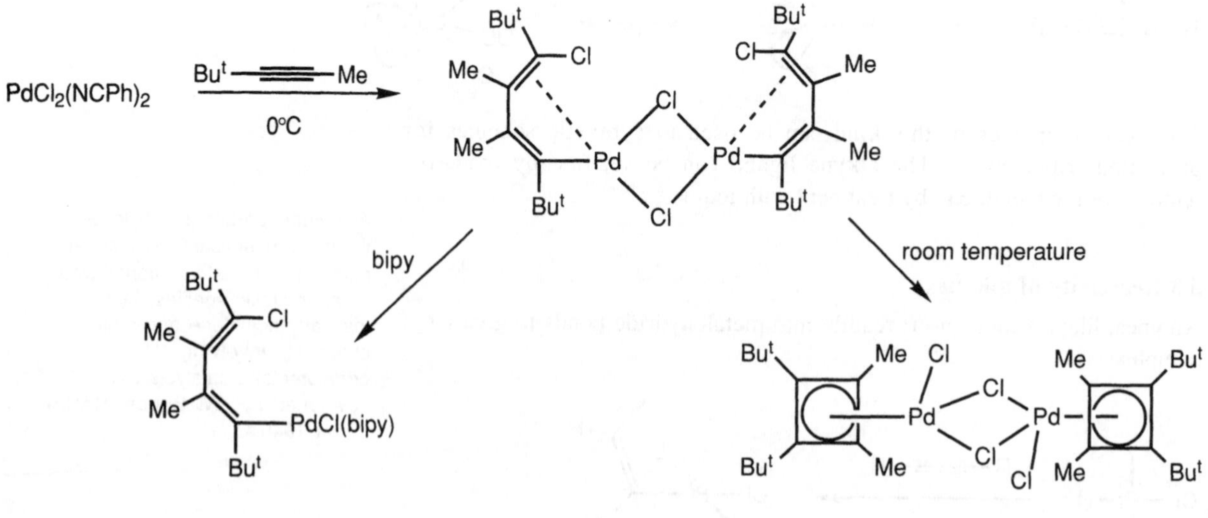

Metal carbonyl clusters react readily with alkynes, and the formation and structure of $Co_2(CO)_6(C_2R_2)$ has been mentioned above. There are numerous similar examples. As a rule the C–C bond order in such clusters is substantially less than three. In some cases C–C bond scission is observed:

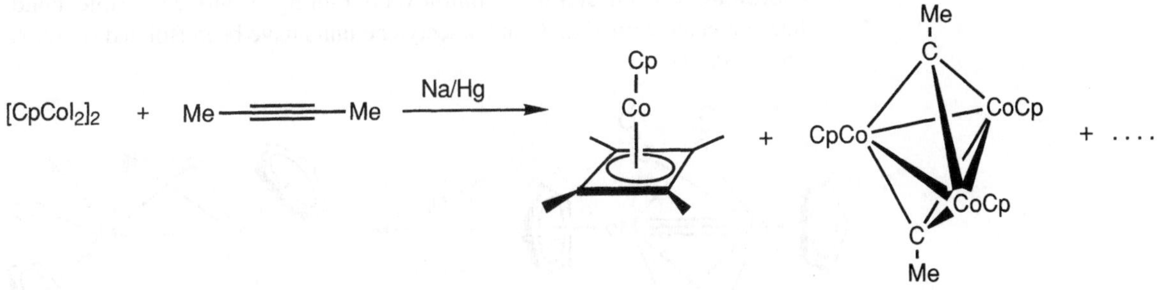

Many transition metal complexes, such as Ni and Pd salts, Nb and Ta chlorides, or sandwich compounds such as $CpML_2$ (M = Co or Rh), catalyse the cyclotrimerization of alkynes to give benzene derivatives. This is a very convenient route to highly substituted arenes; the trimerization of diphenylacetylene, for example, gives hexaphenylbenzene. The reaction is thought to proceed stepwise:

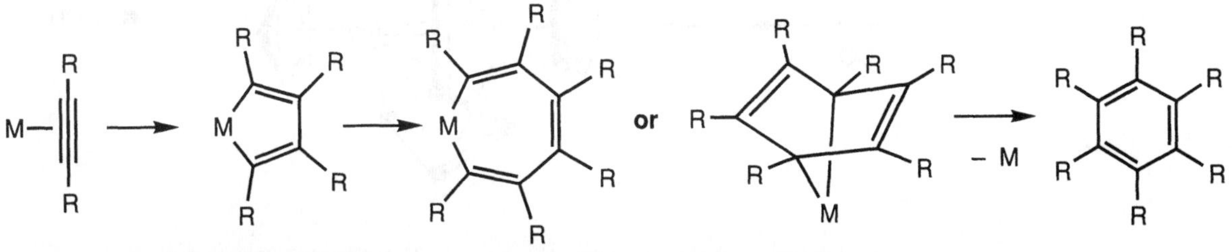

metallacyclopentadiene

The formation of *metallacyclopentadiene* complexes is a very typical alkyne reaction and is frequently observed for a variety of metals. It has been possible in some cases to isolate examples for each of these intermediates, depending on the substituents R:

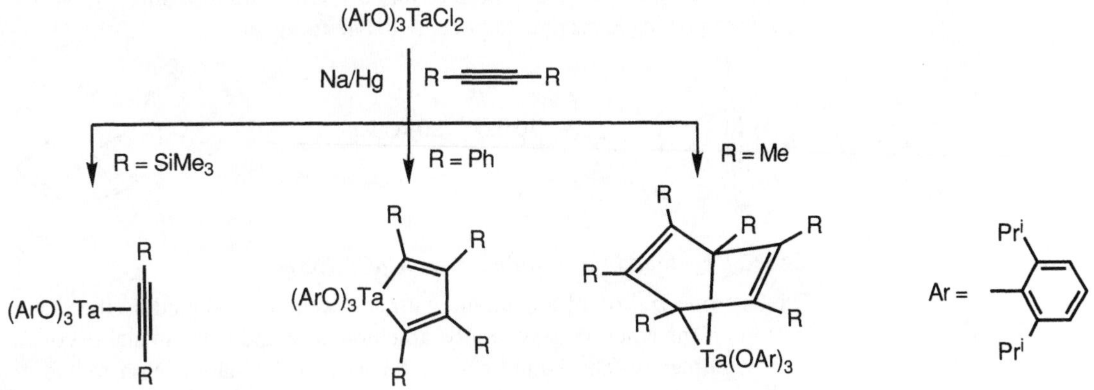

Nickel salts, such as $Ni(CN)_2$, catalyse not only the cyclotrimerization but also the tetramerization of acetylene to cyclooctatetraene.

The controlled stepwise oligomerization of acetylenes within the metal coordination sphere has been observed in the reaction of $[CpMo(CO)_2]_2$, a coordinatively unsaturated complex containing a Mo≡Mo triple bond. Intermediates with 1, 2, 3, and 4 acetylene units have been isolated (F. G. A. Stone, 1978):

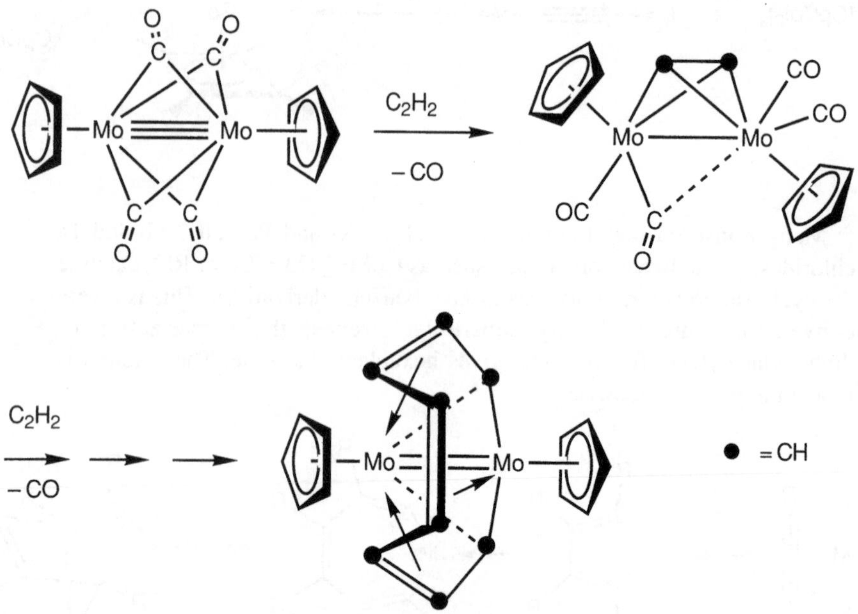

Alkyne complexes in synthesis 1. Cobalt-catalysed alkyne cycloadditions

Pauson–Khand reaction

This reaction allows the coupling of an acetylene, an olefin and CO in one step. It can be made catalytic, with a moderate turnover.

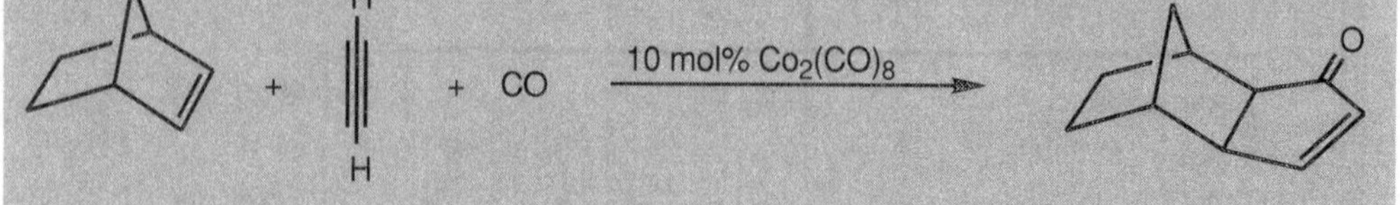

[2+2+2] cycloadditions induced by $CpCo(CO)_2$

The co-trimerization of a dialkyne with an alkene is mediated by the CpCo fragment. The reaction is generally stoichiometric and leads to stable cobalt diene complexes. The ligand can be released by oxidation with iodine or Ce^{IV}.

Similarly, the trimerization of alkynes can be used to synthesize otherwise inaccessible arenes, including oligomeric or polymeric structures:

Ligand control in the cobalt-catalysed pyridine synthesis.

Cobalt complexes of the type $Cp'CoL_2$ catalyse the co-trimerization of acetylenes with nitriles to give substituted pyridines. The reaction is accompanied by the (undesirable) formation of benzene derivatives. Suitable choice of the cyclopentadienyl ligand is used to suppress the formation of arenes and to control the activity of the catalyst.

main products if nitrile concentration is high

side-products

Catalysts containing L = ethylene or 1,5-COD are very much more active than complexes with L = CO, indicating that facile dissociation of L is important and that the reaction is induced by the Cp'Co fragment. For a given L both chemoselectivity and catalytic activity are influenced by the electron donor or acceptor properties of the Cp' ligand. Steric influences are largely absent, and the electronic character of the Cp' ligands is reflected in the ^{59}Co NMR chemical shift which correlates linearly with the activity of the catalyst. ^{59}Co has a very large chemical shift range and is very sensitive to electronic changes in the ligand sphere. Electron-rich Cp' groups such as C_5Me_5 increase the electron density of the metal, indicated by ^{59}Co chemical shifts to high field, resulting in low activity, while electron-withdrawing Cp' ligands activate the catalyst. A similar correlation exists for substituted indenyl ligands.

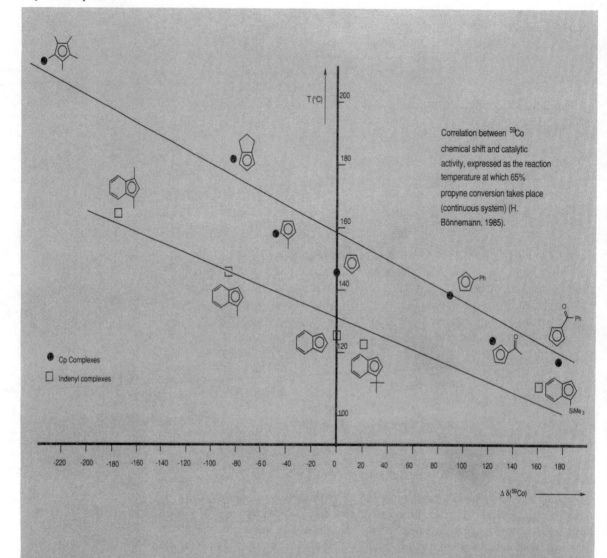

Correlation between ^{59}Co chemical shift and catalytic activity, expressed as the reaction temperature at which 65% propyne conversion takes place (continuous system) (H. Bönnemann, 1985).

Alkyne complexes in synthesis 2. Alkyne polymerization

The polymerization of acetylene is catalysed by $Ti(OBu^n)_4/AlEt_3$ Ziegler–Natta type catalysts and gives fibrillar polyacetylene which, when doped and stretched, has an electric conductivity which exceeds that of copper (H. Naarmann, 1985):

conducting polyacetylene

Carbene complexes of molybdenum also catalyse the polymerization of acetylenes, via a mechanism reminiscent of the ring-opening polymerization of alkenes (*Organometallics 1*, p. 88). Less well-defined catalysts, such as $MoCl_5/SnR_4$ mixtures are also effective.

The distinction between the two mechanisms has been elegantly demonstrated by copolymerizing $PhC{\equiv}CH$ with $Ph^{13}C{\equiv}^{13}CH$ and measuring the $^{13}C{-}^{13}C$ coupling constant of the resulting polymer. The Ti-catalysed polymerization proceeds via insertion of the alkyne into a Ti–C single bond, and the two carbon atoms of the alkyne remain connected via a C=C double bond. The carbene mechanism, on the other hand, gives a product in which the bond order between the carbon atoms of the original alkyne is reduced to 1, leading to a significantly lower $^1J_{C-C}$ coupling constant:

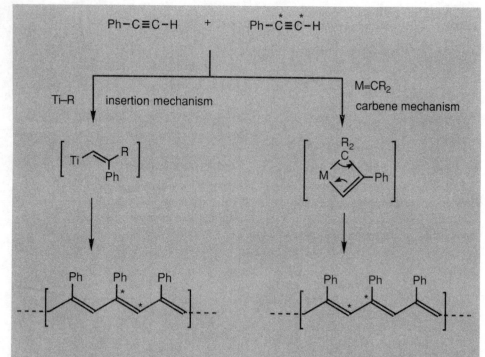

Alkyne complexes in synthesis 3. Benzyne complexes

Benzynes have been mentioned above as highly strained (R–C≡C << 180°!) reactive species which are stabilized by metal coordination. Benzyne complexes can be generated by *ortho*-hydrogen abstraction, and zirconium complexes in particular have found widespread synthetic applications. The intermediate complex, $Cp_2Zr(C_6H_4)$, is unstable but can be isolated as the PMe_3 adduct and undergoes a range of coupling reactions:

alkene, ketone and nitrile insertion products

4 Cyclopentadienyl complexes

In 1948 Miller, Tebboth, and Tremaine, attempting to synthesize amines from olefins and nitrogen in the presence of iron catalysts, found that with cyclopentadiene, an iron-containing compound was formed, $FeC_{10}H_{10}$. Three years later Kealey and Pauson obtained the same compound in an attempt to oxidize C_5H_5MgBr with $FeCl_3$ and, in analogy to main group element alkyls, suggested a structure with two metal–carbon σ-bonds:

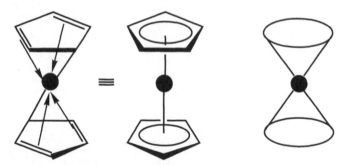

The true nature of this remarkably stable product was recognized independently by G. Wilkinson and R.B. Woodward at Harvard and E. O. Fischer in Munich who suggested a **'sandwich'** or **'double-cone'** structure, with all five C-atoms of a cyclopentadienyl (Cp) ligand interacting with the metal centre:

The new compound, bis-(cyclopentadienyl)iron, was named *'ferrocene'* in the realization that it behaves much like a three-dimensional arene. For example, it undergoes Friedel–Crafts acylations like a benzene derivative:

Soon afterwards a whole series of bis-(cyclopentadienyl)metal complexes Cp_2M were prepared; by analogy these have become known as **metallocenes.**

The salt $K^+C_5H_5^-$ had already been prepared in 1901, and $C_5H_5^-$ was known to be a planar, delocalized system, isoelectronic to benzene (6π-Hückel aromatic):

Since Cp^- carries a negative charge it displaces halide ligands readily from metal salts and forms numerous thermally stable complexes. As a conjugated unsaturated hydrocarbon it possesses π and π^* orbitals and, like olefins (see Chapter 2), acts both as a π-donor and as a π-acceptor. As a consequence most Cp complexes are essentially **covalent**, with little charge separation between metal and ring ligand. The favourable metallocene structure frequently overrides the requirements of the 18-electron rule: *first-row metallocenes exist with electron counts from 15 to 20.*

The discovery of the **sandwich structure** as a bonding principle is one of the cornerstones of organometallic chemistry. As the name suggests, sandwich complexes are compounds in which the metal is intercalated between two hydrocarbon ligands $C_nH_n^{x-}$ with planar, conjugated π-systems. The ligands are predominantly cyclic, although the 'sandwich' bonding principle applies equally to open enyl and diene complexes (see allyl and dienyl complexes, Chapter 5).

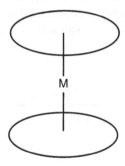

The ring ligands are Hückel aromatic systems. Note that *most transition metal sandwich complexes follow the 18-electron rule*. However, there are notable exceptions, such as the already mentioned 'metallocenes', and lanthanide and actinide complexes of which $U(C_8H_8)_2$ ('*uranocene*') is the best known—the 18-electron rule does not apply to these metals.

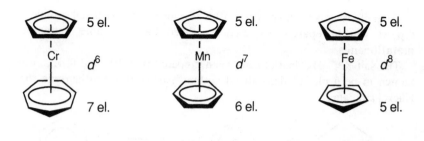

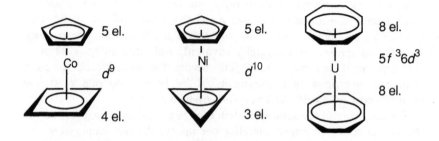

4.1 Synthesis of metallocenes

Monomeric cyclopentadiene is generated by the thermal cracking of the Diels–Alder dimer (dicyclopentadiene) as a low-boiling liquid. The compound is C–H acidic and reacts readily with strong bases such as alkali metals or alkali metal hydrides under deprotonation, to give salts of the cyclopentadienyl anion, $C_5H_5^-$. Alkali metal Cp compounds are essentially ionic.

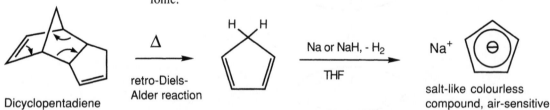

The reaction of the Li, Na or K derivatives with transition metal salts is the most versatile synthetic method for metallocenes:

$$MCl_2 + 2\ Na^+Cp^- \xrightarrow[-\ NaCl]{} MCp_2 \qquad (M = V, Cr, Mn, Fe, Co)$$

$$Ni(acac)_2 + 2\ CpMgBr \xrightarrow[-\ MgBr(acac)]{} NiCp_2$$

It is sometimes possible (and certainly more convenient) to generate the cyclopentadienyl anion *in-situ* by the addition of base to a mixture of the metal salt and cyclopentadiene:

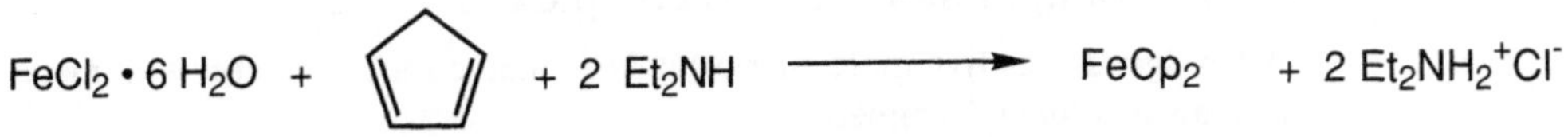

$$FeCl_2 \cdot 6\,H_2O \; + \; \text{(cyclopentadiene)} \; + \; 2\,Et_2NH \longrightarrow FeCp_2 \; + \; 2\,Et_2NH_2{}^+Cl^-$$

The introduction of substituents into the C_5H_5 ring is facile and provides a convenient way to modify the electronic and steric properties of Cp ligands. Pentamethylcyclopentadienyl (C_5Me_5=Cp*) for example greatly increases steric hindrance (it has roughly twice the diameter of C_5H_5), and due to the +I effect of the five methyl groups it is also a very much better electron donor, increasing the electron density of the metal centre and shifting the redox potential of the corresponding metallocene to negative values (i.e. Cp* complexes are more easily oxidized than C_5H_5 analogues). Some widely used cyclopentadienyl ligands are:

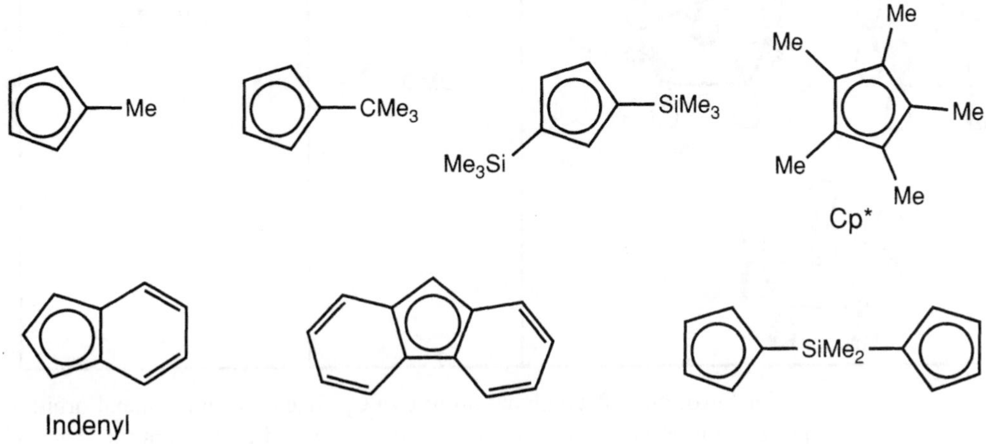

Cp*

Indenyl

4.2 Bonding in metallocene complexes

A Cp ligands has five molecular orbitals available for interaction with metal orbitals of suitable symmetry:

		Symmetry	Metal orbitals
		e_2 LUMO	$d_{xy}, d_{x^2-y^2}$
		e_1 HOMO	p_x, p_y, d_{xy}, d_{xz}
		a_1	d_{z^2}

For ferrocene, the combination of two Cp ligands with the metal orbitals gives the following MO diagram. The HOMOs and LUMOs in this scheme have essentially d-character. Note that the HOMO in the case of Fe is a doubly occupied a_{1g} level, while the LUMOs are degenerate. This scheme is of course applicable to other metallocenes of D_{5d} symmetry and explains, for example, the presence of two unpaired electrons in the case of nickelocene.

The MO diagram illustrates a bonding pattern similar to that for CO complexes, with σ-donor bonds [Cp(a_1)→M($4s$, $3p_z$)], π-donor bonds [Cp(e_1)→M($d_{xz,yz}$, $p_{x,y}$)] and π-acceptor bonds [Cp(e2)← M($d_{x^2-y^2}$, d_{xy})].

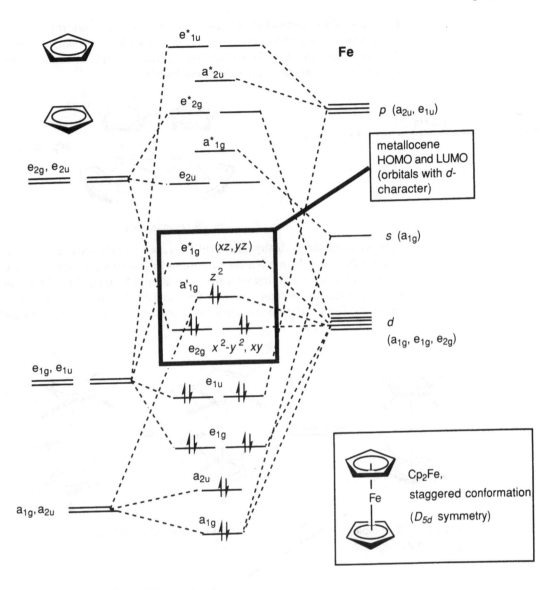

Cp$_2$Fe, staggered conformation (D_{5d} symmetry)

4.3 Properties of metallocenes

Compound	Colour	m.p. (°C)	No. of electrons (unpaired)	Magnetic moment μ_{eff} (B.M.) spin-only value	found
VCp$_2$	purple	167	15 (3)	3.87	3.84
CrCp$_2$	red	173	16 (2)	2.83	3.20
MnCp$_2$	amber/pink	173	17 (5)	5.92	5.81
FeCp$_2$	orange	173	18 (0)	0	0.00
CoCp$_2$	black-purple	174	19 (1)	1.73	1.76
NiCp$_2$	green	173	20 (2)	2.83	2.86

Titanium. Some particularly electron-deficient metals do not form simple metallocenes but seek other ways to increase their electron count, as well as their oxidation state. Cp_2Ti does not exist but reacts with C–H bonds to give hydrido species; several isomers have been found:

$$Cp_2TiCl_2 \xrightarrow{\text{Na/Hg}} \text{"}Cp_2Ti\text{"} \equiv$$

and isomers

16 VE, diamagnetic

By contrast, the sterically hindered $Cp*_2Ti$ is relatively stable at 0°C; it is monomeric and at higher temperature forms an equilibrium with a hydrido alkyl species. Such an intramolecular 'activation' of a methyl C–H bond of $Cp*$ is not unusual for complexes of Ti^{II} and Ti^{III}: $Cp*$ is not necessarily a 'spectator ligand'. The electron-deficient character is further illustrated by the formation of N_2 complexes:

K

N_2

Isolable 14 VE complex, yellow, paramagnetic

16 VE, green, diamagnetic

$$Cp*_2Ti\!-\!N\!\equiv\!N\!-\!TiCp*_2$$

Vanadium. In contrast to Cp_2Ti, the highly air-sensitive 15 VE complex Cp_2V is isolable. Its reactions demonstrate its electron-deficient character and the tendency to reach higher oxidation states. Even adducts between Cp_2V and neutral ligands such as alkynes are best described as oxidative addition products (metallacyclopropenes) due to extensive back-bonding:

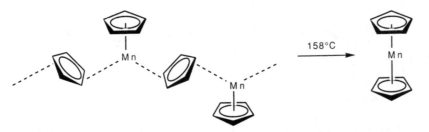

R = COOMe:

143.5°

1.27 Å

Cp$_2$V

VCl$_3$(THF)$_3$ or
VCl$_2$(THF)$_2$ → NaCp → Cp$_2$V

R≡R → Cp$_2$V

R'X → Cp$_2$VR' + Cp$_2$VX

CS$_2$ → Cp$_2$V

Manganese. Manganese(II), as a high-spin d^5 ion, has zero crystal field stabilization energy and behaves as a 'hard' ion, similar to Zn^{2+} or Mg^{2+}. Consequently, the Cp ligand in MnCp$_2$ is *less covalently bonded than in other metallocenes*, and MnCp$_2$ is unusual in the transition metal series in adopting a **polymeric structure** similar to that of ZnCp$_2$. As a further expression of the ionic character of MnCp$_2$, the compound forms 19- to 21-electron adducts with donor ligands, such as Cp$_2$Mn(THF)$_2$, Cp$_2$Mn(PMe$_3$) and Cp$_2$Mn(dmpe). Cp$_2$Mn exists in two phases: at 158°C the amber polymeric low-temperature modification gives way to the pink monomeric metallocene structure:

158°C →

Mn Mn Mn

Mn

Linear polymeric 'multidecker' structures have been found for lithium Cp derivatives, e.g. [Li(η^5-C$_5$H$_4$SiMe$_3$)]$_\infty$:

The magnetic behaviour of Cp$_2$Mn is unusual. It is high-spin (S = 5/2, μ_{eff} = 5.9 B.M.) at room temperature but close to the high-spin/low-spin crossover point; the energy difference is only *c.* 2 kJ mol^{-1} and small intermolecular forces in frozen solutions are sufficient to enforce the low-spin configuration. Because of the higher field strength of Cp*, Cp*$_2$Mn is low-spin (S = 1/2, μ_{eff} = 2.18 B.M.); this monomeric complex has metallocene structure and does not form a polymeric modification.

Iron. With 18 valence electrons, ferrocene is the most stable member in the metallocene series. It sublimes readily and is not attacked by air or water but can be oxidized reversibly, electrochemically or by oxidising reagents such as iodine, to give the blue ferricenium cation, [Cp$_2$Fe]$^+$. The number of ferrocene derivatives is vast:

R

Li

R

Li

R

Li

R = SiMe$_3$

Ferrocene may be regarded as a fairly *electron rich arene*; in Friedel-Crafts acylations it reacts *c.* 10^6 times faster than benzene. It forms deep green charge transfer complexes with electron acceptors such as tetracyanoethylene, $Cp_2Fe\cdot(TCNE)$.

The oxidation of orange Cp_2Fe to the deep blue, paramagnetic ion Cp_2Fe^+ ($E° = +0.31$ V vs. standard calomel electrode) is reversible; Cp_2Fe/Cp_2Fe^+ couples are used in electrochemistry. Binuclear ferrocene derivatives allow the determination of electron transfer rates in mixed-valence compounds (e.g. by Mößbauer spectroscopy).

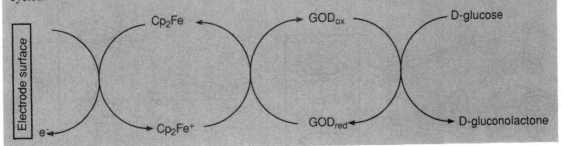

Cp$_2$Fe / Cp$_2$Fe$^+$ couples in biosensors

The oxidation of glucose by the enzyme glucose oxidase (GOD) to gluconolactone is sensitive and specific, and therefore suitable for the determination of glucose with sensors. The direct electrochemical measurement of the amount of oxidized product is however impeded since the enzyme does not react with electrode surfaces directly. A redox-active mediator is required in order to facilitate the quantitative oxidation of glucose under catalytic conditions. Ferrocene acts as such a mediator, enabling the electrochemical determination of glucose via a series of connected redox cycles:

Cobalt. The 19 VE complex Cp$_2$Co has one unpaired electron and, consequently, acts as a powerful *reducing agent* (E° = -0.90 V vs. SCE). It is stable in an inert atmosphere and sublimes readily but is highly air-sensitive. Oxidation gives yellow Cp$_2$Co$^+$ which is more stable than ferrocene and unaffected by strong aqueous acids and bases. Cp$_2$Co$^+$ has the diameter of a large alkali cation, comparable to Cs$^+$.

$$[Cp_2Co]^+X^- \xleftarrow{RX} Cp_2Co \xrightarrow[H_2O]{O_2} [Cp_2Co]^+OH^-$$

Cobaltocene intercalates into inorganic solids with layer structures, for example SnS$_2$ or TaS$_2$, to give compounds such as [SnS$_2$(Cp$_2$Co)$_{0.31}$]. Materials of this kind are of interest because of their magnetic and electronic properties; for example, intercalation raises the temperature at which superconductivity is observed. Recent X-ray results indicate that the Cp$_2$Co molecules are oriented parallel to the layers of the host lattice:

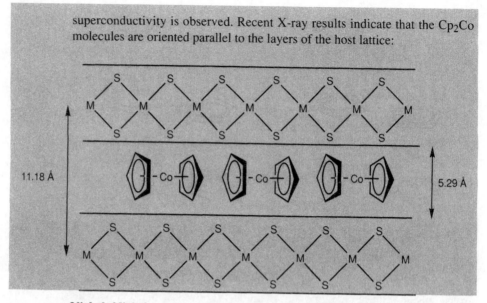

Nickel. Nickelocene is a rare example of a 20 VE complex; it contains two unpaired electrons (see MO diagram of metallocenes, p. 47). The reactivity of nickelocene reflects the tendency of Ni to attain an 18-electron configuration. Note the unusual formation of the tripledecker sandwich complex [Ni$_2$Cp$_3$]$^+$ by capping Cp$_2$Ni with a CpNi$^+$ fragment—the first complex of this kind.

Metallocenes of second and third row metals. Few heavy transition metals form metallocene sandwich structures, and the use of the bulky Cp* ligand is often required to obtain isolable products. Reduction of Cp*$_2$ZrCl$_2$ gives highly reactive 'Cp*$_2$Zr' which requires stabilization by donor ligands:

$$Cp*_2ZrCl_2 \xrightarrow{Na/Hg} \boxed{\text{"}Cp*_2Zr\text{"}}$$

$$\xrightarrow{C_2H_4} Cp*_2Zr-\|$$

$$\xrightarrow{H_2} Cp*_2ZrH_2$$

$$\xrightarrow{N_2} Cp*_2Zr-N\equiv N\cdot ZrCp*_2$$

$$\downarrow N_2$$

$$Cp*_2\underset{N_2}{Zr}-N\equiv N\overset{N_2}{ZrCp*_2}$$

'NbCp$_2$', 'MoCp$_2$' and 'WCp$_2$' adopt structures related to those described for 'TiCp$_2$', i.e. they form dinuclear hydrido species via activation of a C–H bond of Cp. Cp$_2$Re is unstable but can be generated at low temperatures, while Cp*$_2$Re is isolable. Only Ru and Os form stable ferrocene analogues. While it has long been known that ferrocene reacts with Hg(OAc)$_2$ to give a mono–mercuration product, it was discovered recently that the mixed sandwich RuCpCp* allows the substitution of all five hydrogens of the Cp ring by mercury:

$$Cp*RuCp + 5\ Hg(OAc)_2 \longrightarrow Cp*Ru[C_5(HgOAc)_5] + 5\ HOAc$$

[Cp*Ru(NCMe)$_3$]$^+$ reacts with RuCp*$_2$ to give the tripledecker [Ru$_2$Cp*$_3$]$^+$ (see multidecker complexes, Chapter 6). 'RhCp$_2$' is an unstable radical, CpRh(η^4-C$_5$H$_5$), observed only by matrix isolation, and dimerizes rapidly on warming. There are no Ir, Pd and Pt metallocenes.

4.4 Bent sandwich complexes

The presence of ligands other than cyclopentadienyl in a metallocene complex enforces a geometry in which the Cp ligands are no longer co-planar: they adopt a 'bent sandwich' structure. In such complexes the firmly bonded Cp ligands may be regarded as *protecting groups* while chemical transformations involve only the frontier orbitals of the molecule, i.e. those oriented towards the non-Cp ligands. The protective function of Cp becomes evident for example when the hydrolytic stability and low Lewis acidity of Cp$_2$TiCl$_2$ is compared with the vigorous hydrolysis of the strong Lewis acid TiCl$_4$.

Because of the 18-electron rule bent metallocene complexes are restricted to metals with a low number of *d*-electrons. Some examples:

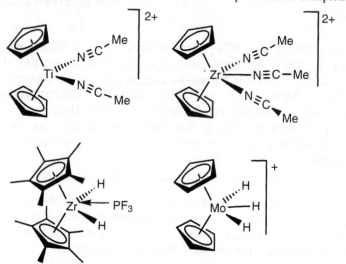

d^0, 14 el.　　d^2, 18 el.　　d^1, 17 el.　　d^4, 18 el.　　d^4, 18 el.

4.4.1 Bonding in bent sandwich complexes

The nature of the 'business end' of bent metallocenes, i.e. the shape and number of the frontier orbitals, becomes evident by comparing the structures and the Lewis acid/Lewis base character of some representative complexes:

d^0-Complexes such as [Cp$_2$M]$^{2+}$ (M = Ti, Zr, Hf) are able to coodinate up to three donor ligands (for Ti with its small ionic radius the number is restricted to two) while Cp*$_2$ZrH$_2$ accepts only one. By contrast, the d^2 complex Cp$_2$MoH$_2$ possesses an electron pair in addition to the two hydride ligands; it acts as a base and can be protonated. Note that all the ligands are arranged in the 'equatorial' plane of the molecules, i.e. the plane bisecting the Cp-M-Cp angle.

These observations agree with MO calculations which show the presence of three frontier orbitals whose occupancy depends on the *d*-electron count of the metal. The orbitals denoted $d_{\sigma}1$ and $d_{\sigma}2$ are involved in σ-bonds e.g. to H ligands in Cp$_2$MH$_2$. In Cp$_2$ZrH$_2$ d_{π} is the LUMO, while in Cp$_2$MoH$_2$ d_{π} is occupied by two *d*-electrons.

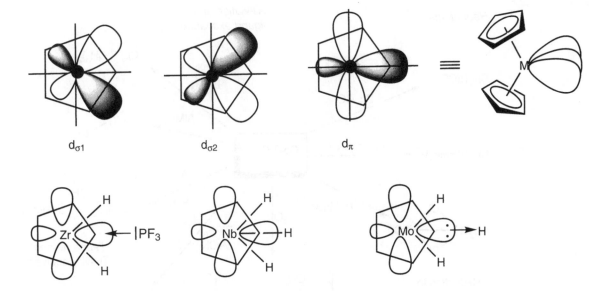

$d_{\sigma 1}$ $d_{\sigma 2}$ d_{π}

Metallocene halides and hydrides

Metallocene dihalides are typical examples of bent sandwich complexes and are ideal starting materials for ligand exchange and redox reactions. They are commonly synthesized from the corresponding metal halides:

$$MCl_4 + 2\ NaCp \xrightarrow{\text{THF, } -2\ NaCl}$$

ca. 130°   ca. 95°

M = Ti, Zr, Hf,
V, Nb, Ta,
Mo, W

The reaction pattern of metallocene dihalides is illustrated below for reactions of Cp_2TiCl_2. Note that only the non-Cp ligands participate. The reaction types observed are:

- alkylations (substitution of Cl by σ-alkyl ligand);
- reductions;
- exchange reactions with donor ligands.

Potential organometallic pharmaceuticals: apart from their synthetic potential, Cp_2TiCl_2 and Cp_2VCl_2 have attracted interest because of their *in vitro* activity against certain types of cancer cells, possibly because the Cp_2M^{2+} fragment coordinates to DNA similar to Pt^{II} anti-cancer drugs.

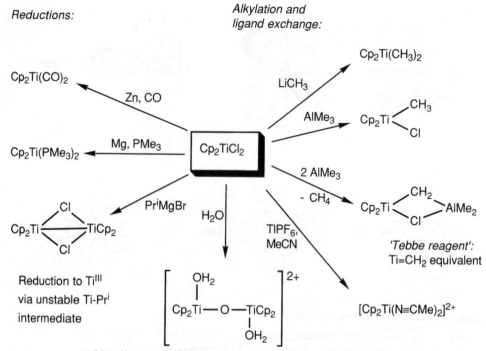

Reductions:

Alkylation and ligand exchange:

Reduction to Ti^{III}
via unstable Ti-Pr^i
intermediate

'Tebbe reagent':
Ti=CH$_2$ equivalent

Metallocene hydrides are formed predominantly by second and third row transition metals:

d^0	d^2	d^4
[Cp*$_2$ScH]$_n$,Cp*$_2$ScH(THF)	Cp$_2$MH(CO) (M = Nb,Ta)	[Cp$_2$M(H)Li]$_4$ M = Mo, W)
[Cp$_2$Zr(H)Cl]$_n$,	Cp$_2$MH$_2$ (M = Mo, W)	Cp$_2$MH (M = Tc, Re),
[(C$_5$H$_4$R)$_2$M(μ-H)H]$_2$	[Cp$_2$MoH(CO)]$^+$,	
(M = Zr, Hf; R = Me, Bu^t, SiMe$_3$)	[Cp$_2$WH(C$_2$H$_4$)]$^+$,	
Cp*$_2$MH$_2$ (M = Ti, Zr, Hf)	[Cp$_2$ReH$_2$]$^+$	
Cp$_2$MH$_3$ (M = Nb, Ta)		
[Cp$_2$MH$_3$]$^+$ (M = Mo, W)		

The first of this class to be isolated was Cp$_2$ReH:

$$ReCl_5 + 2\ NaCp \xrightarrow[\text{THF}]{\text{NaBH}_4} Cp_2ReH$$

$\delta(^1H)$ = 3.64 (Cp), −23.5 ppm (Re-H)

NMR is frequently useful for the identification of hydrido complexes: in most (but not all) cases the ^1H NMR resonance of the hydrido ligand is observed high-field of TMS.

In agreement with the MO considerations for bent metallocenes outlined above, Cp$_2$ReH (d^4) possesses two non-bonding electron pairs and acts as a base, in strength comparable to an amine. Some of its reactions are outlined below:

Tailored metallocenes and the stereospecific polymerization of 1-alkenes

Mixtures of metallocene dihalides of group IV metals (Ti, Zr, or Hf) and aluminium alkyls as activators are highly active catalysts for the polymerization of olefins (Ziegler–Natta polymerization; see *Organometallics 1*, p. 69 ff.). The active species is a coordinatively unsaturated cationic alkyl complex, $[Cp_2M-R]^+$:

The polymerization of 1-alkenes such as propene with such a catalyst would lead to a polymer with randomly oriented alkyl side-chains. However, since cyclopentadienyl ligands are easily modified by substituents and can be tailored to suit particular requirements, appropriate ligand design has been used to control the orientation of the coordinated propene monomer in such a way that either isotactic or syndiotactic polymers must result. Such stereoregular polymers have quite different physical properties, such as much higher melting points and greater toughness. For example, in bis-(indenyl) complexes the rotation of the ligands can be prevented by connecting them via a $-CH_2-CH_2-$ bridge, to give a rigid and stereoselective ligand framework (a so-called *ansa*–metallocene, from *ansa* = handle). Catalysts containing this ligand orient the methyl group of the incoming propene monomer away from the indenyl ligand prior to the insertion step to give highly isotactic polypropene:

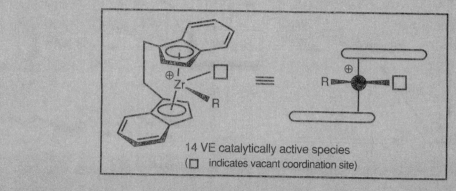

14 VE catalytically active species
(☐ indicates vacant coordination site)

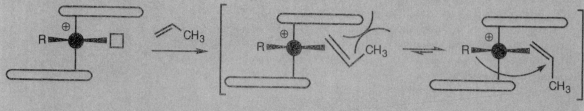

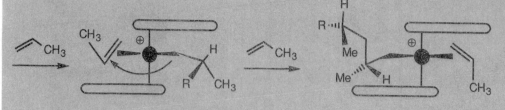

isotactic polypropene

Similarly, a ligand system in which a large, raft-like Cp analogue such as fluorenyl is combined with a small C_5H_5 ligand leads, via an analogous mechanism, to the chain growth of syndiotactic polypropene.

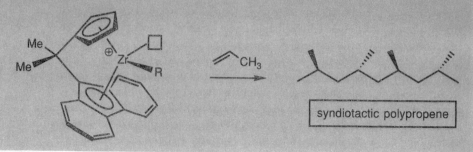

syndiotactic polypropene

4.5 Cyclopentadienyl as a non-spectator ligand: ring slippage and fluxionality

In all the examples described so far, the Cp ligand was always η^5 (pentahapto) coordinated and could be regarded as a firmly bound 'protecting group', i.e. a spectator ligand which does not participate in chemical transformations. Such a ligand can be thought to occupy three coordination sites, e.g. $CpFe(CO)_2Cl$ and $FeCp_2$ can be regarded as octahedral complexes:

Although this form of Cp bonding is very common and very stable, cyclopentadienyl ligands are far from rigid and able to adopt alternative coordination modes if required. Several types of Cp mobility has been observed:

☞ Cp ring rotation (ring whizz);
☞ η^1-Cp and σ-bond shift;
☞ η^5–η^1 interchange;
☞ η^5–η^3 ring slippage.

η^5-Cp ligands rotate very rapidly around the metal–Cp (centroid) axis, giving rise to the typical sharp Cp singlets in the ^{1}H and ^{13}C NMR spectra, a feature of high diagnostic value in complex characterization. Although in the solid state the two Cp ligands in a metallocene complex may adopt staggered or eclipsed conformations, the rotational barrier which the Cp ligands experience is very small (c. 8–9 kJ mol^{-1}). It is practically impossible to 'freeze out' the C_5H_5 rotation. Very recently a rare case of hindered rotation of a Cp ligand has been observed: cold ($-100°C$) solutions of the complexes $[CpM(styrene)(PPh_3)_2]^+$ (M = Ru, or Os) give ^{13}C NMR spectra in which the Cp signal is split into five components, as expected since the lack of symmetry would render all five CH groups of the Cp ligand inequivalent if Cp rotation was slow on the NMR time scale (R. Mynott, 1990).

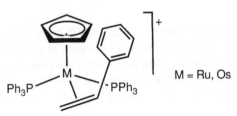

M = Ru, Os

In contrast to transition metals, η^1-Cp bonding is common with main group elements, e.g. $Hg(\eta^1$-$Cp)_2$. Both η^1 and η^5 ligands exist in

$CpFe(CO)_2(\eta^1-C_5H_5)$. In both cases the η^1-Cp ligand is fluxional and undergoes rapid 1,2-sigmatropic shifts of the metal–carbon σ-bond, so that above room temperature all five carbons are in effect identical on the NMR time scale. For example, at 30°C the ^1H NMR spectrum of the Fe complex consists of two sharp singlets, while on cooling one of these becomes the multiplet expected for a 'frozen–out' η^1-Cp.

$CpFe(CO)_2(\eta^1-C_5H_5)$ was the first organometallic compound for which fluxional behaviour was detected. Since then fluxionality has become recognized as one of the most interesting and important aspects of organometallic complexes. Another classic example is $TiCp_4$. For electronic as well as steric reasons it is clear that all the Cp ligands cannot be η^5-bonded to titanium at the same time. In contrast to $CpFe(CO)_2(\eta^1-C_5H_5)$ all four Cp ligands are involved in the exchange process:

By contrast, the much larger U^{4+} ion is able to bind four η^5-Cp ligands: UCp_4 is not fluxional.

While molybdenum forms a stable 18 VE complex $Cp_2Mo(CO)$, the tungsten analogue reacts with excess CO to give an apparent 20 VE species, $Cp_2W(CO)_2$. Although the NMR spectrum shows that both Cp ligands are identical in solution, the solid-state structure confirms that one Cp ring is only three-coordinate ($\eta^5 \rightarrow \eta^3$ ring slippage): the 18 VE electron count in maintained.

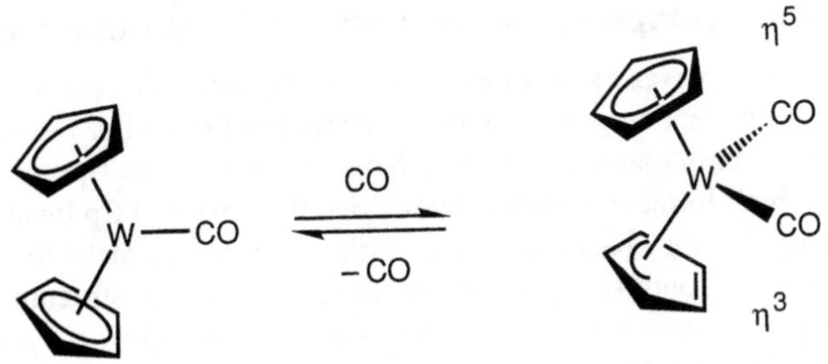

Such an $\eta^5 \rightarrow \eta^3$ rearrangement is more common for indenyl ligands where η^3 coordination is facilitated by the gain in aromaticity in the six-membered ring:

Such rearrangements can have important consequences for the reactivity of metal complexes. For example, substitution of ethylene by other donor ligands in the 18 VE complex $CpRh(C_2H_4)_2$ is slow, whereas the indenyl analogue reacts instantaneously:

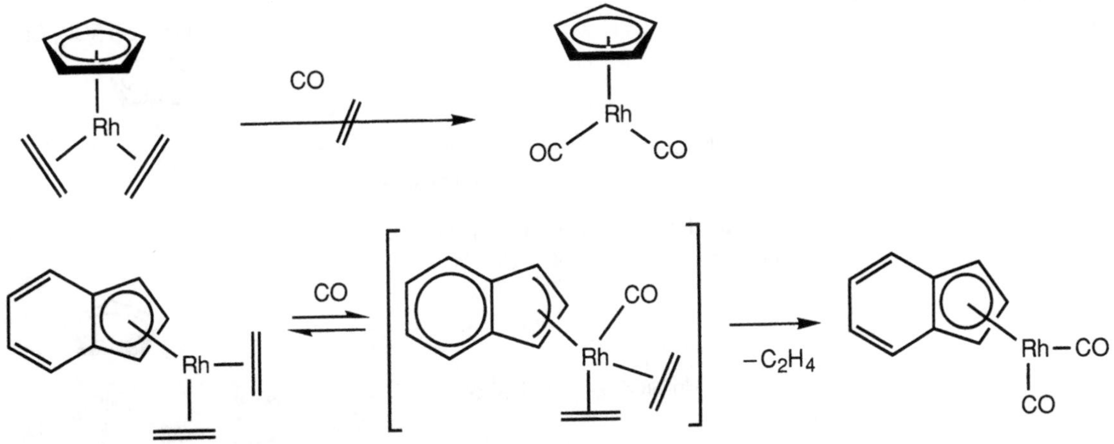

Whereas in the indenyl case the $\eta^5 \rightarrow \eta^3$ ring slippage is reversible, strong donor ligands such as PMe_3 may, in exceptional circumstances, be able to displace Cp ligands to give isolable η^3 and η^1 species:

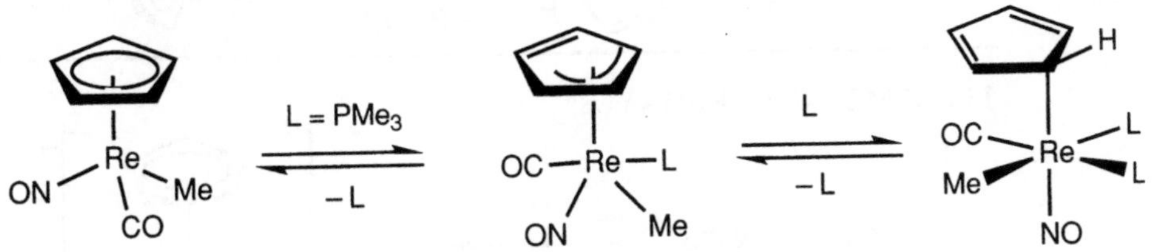

4.6 Mono-cyclopentadienyl (*half-sandwich*) complexes

Mono-Cp complexes exist for every transition metal. The combination of a Cp ligand ('protecting group') with a wide range of σ- and/or π-donor ligands has proved particularly successful. Note that numerous Cp complexes of metals in high formal oxidation states exist: the (anionic) Cp ligand does not necessarily require a back-bonding contribution from the metal to form stable complexes (in contrast e.g. to arenes and alkenes). The structures of these mono-Cp or 'half-sandwich' complexes is aptly described as *'piano stool'* geometry. Monomeric and metal–metal bonded dinuclear compounds exist (as required by the 18-electron rule), as well as numerous structurally related arene derivatives (see Chapter 6).

Synthesis of mono-Cp complexes

FROM METAL SALTS:

Unlike NaCp, TlCp is an air- and hydrolysis-stable reagent, ideal for the introduction of Cp ligands on a small scale. TlCp is prepared from Tl_2SO_4 and CpH on addition of aqueous KOH and purified by sublimation to give pale-yellow fibrous crystals.

similar: $CpTiCl_3$, $CpTaCl_4$

Due to the electron-donor properties of PMe_3, $CpCo(PMe_3)_2$ is a strong metal base, unlike $CpCo(CO)_2$.

FROM METAL CARBONYLS:

NiCp$_2$ + Ni(CO)$_4$ $\xrightarrow{- CO}$

BY OXIDATION:

[CpMo(CO)$_3$]$_2$ $\xrightarrow[\substack{CHCl_3 \text{ / trace } H_2O, \\ \text{light}}]{O_2}$

Cp*Re(CO)$_3$ $\xrightarrow[H_2O]{H_2O_2}$

Structures of half-sandwich complexes

Cyclopentadienyl metal carbonyls and their derivatives are the most numerous half-sandwich complexes. There is a continuous series of isoelectronic mono- and dinuclear compounds; they obey the 18-electron rule. Single, double, and triple metal–metal bonds are formed if necessary in order to achieve this electron count. All these complexes are useful starting materials and are widely used in transition metal mediated synthesis and in catalysis. The metal–metal bonded compounds are particularly reactive.

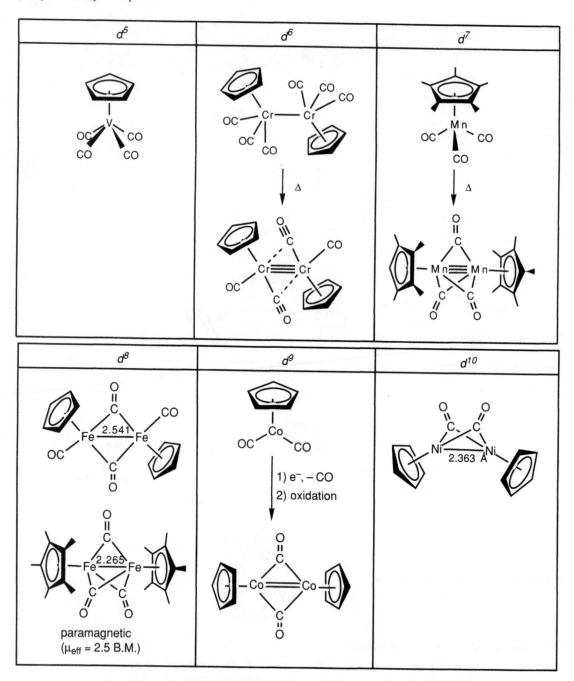

[CpFe(CO)$_2$]$_2$, the most important member of this group by virtue of its stability and widespread use in synthesis, exists in solution as a mixture of isomers with and without CO bridges; the Cp groups can be mutually *cis* or *trans*. The Fe–Fe bond is easily split, and the compound is the parent of numerous halide, hydride, and alkyl derivatives:

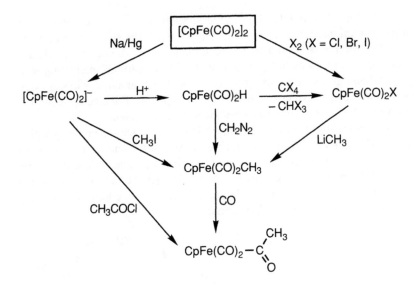

5 Allyl and dienyl complexes

The allyl ligand (CH_2CHCH_2) is the simplest in a series of non-cyclic conjugated anionic (enyl) ligands. Two types exist: the η^1-allyl group is σ-bonded to the metal, with a free C=C double bond, whereas η^3-allyls are bonded in a similar way to Cp ligands with all three carbons interacting with the metal and give complexes with structures closely analogous to sandwich compounds such as cyclopentadienyls (see Chapter 4). For electron counting purposes, η^3-allyl is regarded as a neutral 3-electron ligand.

Coordinated allyl is susceptible to nucleophilic attack—a property which has made allyl complexes invaluable in synthesis and is the basis of numerous catalytic reactions. The site of nucleophilic attack depends on the nature of the metal: allyl groups with cationic character are attacked in the 1,3-position (the usual case), while anionic allyls (e.g. complexes of early transition metals) react in the 2-position:

5.1 Synthesis of allyl complexes

Synthesis from allylic halides

Allylic halides or carboxylates undergo facile oxidative addition reactions with low-valent metal complexes such as metal carbonyl anions, or with zerovalent metals such as nickel or palladium. It is synthetically convenient to prepare zerovalent complexes *in situ*, as demonstrated by the palladium reaction where CO is used as the reducing agent.

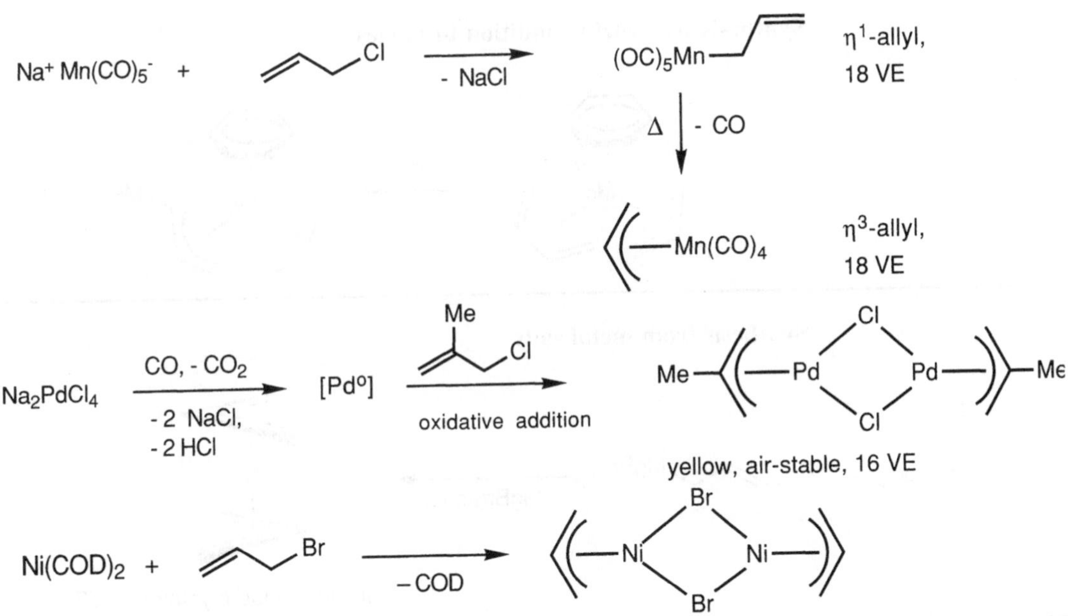

$Na^+ Mn(CO)_5^-$ + [allyl chloride] $\xrightarrow[- NaCl]{}$ $(OC)_5Mn$—[allyl] η^1-allyl, 18 VE

$\Delta \downarrow$ - CO

[allyl]—$Mn(CO)_4$ η^3-allyl, 18 VE

Na_2PdCl_4 $\xrightarrow[\substack{- 2\,NaCl, \\ - 2\,HCl}]{CO, - CO_2}$ $[Pd^0]$ $\xrightarrow[\text{oxidative addition}]{\text{Me—[methallyl chloride]}}$ Me—[allyl]—Pd⟨Cl⟩⟨Cl⟩Pd—[allyl]—Me

yellow, air-stable, 16 VE

$Ni(COD)_2$ + [allyl bromide] $\xrightarrow[- COD]{}$ [allyl]—Ni⟨Br⟩⟨Br⟩Ni—[allyl]

Synthesis from alkenes

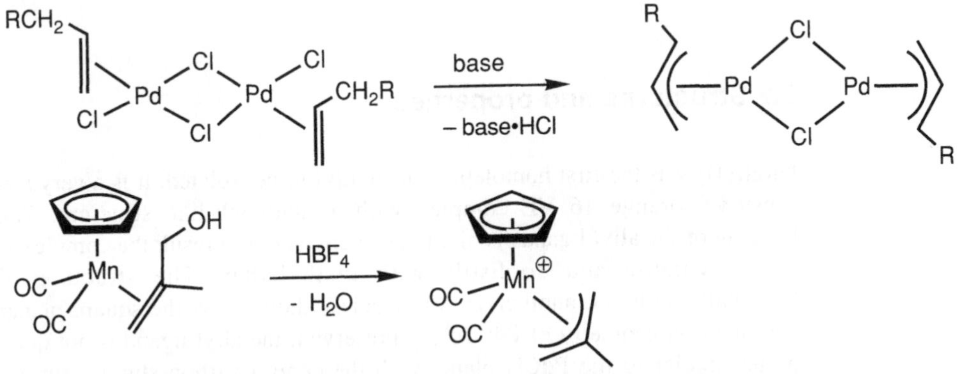

RCH_2[complex] $\xrightarrow[- base \cdot HCl]{base}$ [product]

[Cp Mn(CO)_2 complex with OH] $\xrightarrow[- H_2O]{HBF_4}$ [Cp Mn(CO)_2 cationic complex]

Synthesis from protonation of dienes

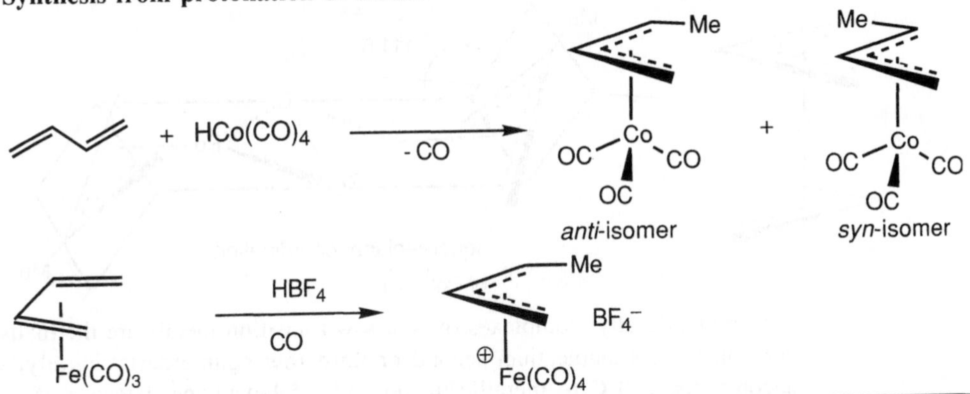

[butadiene] + $HCo(CO)_4$ $\xrightarrow[- CO]{}$ [anti-isomer Co complex] + [syn-isomer Co complex]

anti-isomer   *syn*-isomer

[diene Fe(CO)_3 complex] $\xrightarrow[CO]{HBF_4}$ [cationic Fe(CO)_4 complex] BF_4^-

Synthesis by hydride addition to dienes

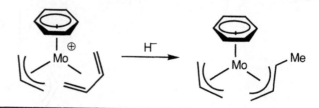

Synthesis from metal salts

$Ni(acac)_2 + 2$ ⟋⟍MgBr $\xrightarrow{- MgBr(acac)}$

volatile orange crystals, 16 VE,
highly air- and temperature sensitive.

$Mo_2(OAc)_4 + 4$ ⟋⟍MgCl $\xrightarrow{- MgCl(OAc)}$ $Mo_2(allyl)_4$

5.2 Structures and properties

Ni(allyl)$_2$ was the first homoleptic metal allyl to be isolated. It is a very air-sensitive, orange 16 VE complex with a sandwich-like structure. The bonding of the allyl ligands is highly covalent, and as a result the complex is highly volatile and co-distils with diethylether. The structure of [(methallyl)PdCl]$_2$, another 16 VE species, shows how the square-planar coordination geometry of Pd^{II} (d^8) is preserved; the allyl ligand is not quite perpendicular to the PdCl$_2$ plane, with the central carbon slightly further from the metal than the two terminal carbons.

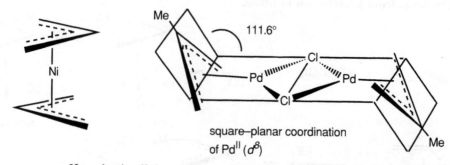

square–planar coordination
of Pd^{II} (d^8)

Homoleptic allyl complexes of first row transition metals are thermally very much less stable than second or third row counterparts. Ni(allyl)$_2$ decomposes > 0°C to metallic nickel and 1,5-hexadiene; because of its lability it serves as a source of zerovalent nickel ('naked nickel') under very smooth conditions. Co(allyl)$_3$ is even more sensitive and decomposes above −55°C. By contrast, Rh(allyl)$_3$ is thermally stable well above room temperature and not affected by air and moisture.

Allyl ligands may be bridging as well as terminal, as in [Pd(allyl)(PPr^i$_3$)]$_2$ or Mo$_2$(allyl)$_4$ where they span the metal–metal bonds: an extension of the sandwich principle to dinuclear species.

$$Pd^0L_2 \ + \ Pd^{II}(\text{methallyl})_2 \ \longrightarrow \ \text{L—Pd} \quad \text{Pd—L}$$

$$L = PPr^i_3$$

5.3 Reactivity of metal allyls

The most prominent characteristic of η^3-allyl ligands is the susceptibility to nucleophilic attack; this is particularly pronounced in cationic complexes. Because of their stability and ease of access, allyl compounds of palladium have found widespread use. Complexes of the type [(allyl)Pd(PR$_3$)$_2$]$^+$ are formed on addition of phosphine and may be generated *in situ*:

Reactions of this kind are frequently carried out catalytically by the addition of Pd(PPh$_3$)$_4$ to mixtures of allylic halides or acetates and a nucleophile.

Allyl ligands can readily undergo $\eta^3 \rightarrow \eta^1$ rearrangements, depending on the electron count of the metal (16/18-electron rule), particularly in the case of electron-rich noble metals. The η^1-allyl ligands behave like alkyl groups and react with electrophiles such as CO$_2$ under insertion:

Changes in hapticity such as the example above are not infrequent in palladium chemistry. In mixed-ligand complexes, ligand-dependent equilibria may be found, as in CpPd(allyl)(L) (L = PPr$^i_3$) where the η^1-Cp/η^3-allyl structure was confirmed crystallographically:

The intramolecular nucleophilic attack on η^3-allyl in electron-rich complexes occurs, as explained above, on C-1 or C-3. If the nucleophile is H$^-$, an olefin complex results; such a process is a model for the allylic mechanism of double bond isomerization in olefins. A temperature-dependent equilibrium between an olefin and an allyl hydride complex that illustrates this process has been established in some cases:

By contrast, nucleophilic attack on C-2 is observed for allyl complexes of early transition metals where the allyl ligand has substantial anionic character:

Reactivity of allyl complexes: butadiene telomerization

Telomerizations are oligomerization reactions which involve the incorporation of the two halves of a reactive molecule A–B at opposite ends of an olefin oligomer. Butadiene telomerizations are catalysed by Pd^0 complexes; they represent a very facile one-step entry into functionalized diene synthesis. There is ample NMR evidence for the course of this reaction; the structures of intermediates such as the Pd η^1,η^3-C$_8$H$_{12}$ complex was determined by X–ray crystallography (L = PMe$_3$).

L = phosphine or phosphite; X = OAc, OMe, NR$_2$, CH$_2$NO$_2$, COOR, SiMe$_3$.

Similar allylic intermediates play an essential role in the cyclooligomerization of butadiene by nickel(0) catalysts. Phosphine-free systems ('naked nickel') give the cyclotrimer, *trans, trans, trans*-cyclododecatriene. These nickel reactions belong to the best understood homogeneously catalysed systems to date (G. Wilke, since 1955).

The trimer, 1,5,9-cyclododecatriene, is produced industrially using a titanium catalyst, via a similar catalytic cycle; it is a precursor for nylon-12.

If phosphines or phosphites are present, the trimerization is suppressed and cyclodimers are produced. The product selectivity is controlled by the concentration and nature of the ligand (see electronic and steric parameters of phosphines, *Organometallics 1*, p. 14 ff.).

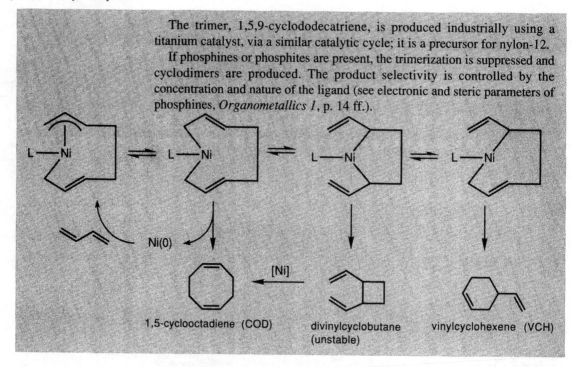

1,5-cyclooctadiene (COD)

divinylcyclobutane
(unstable)

vinylcyclohexene (VCH)

5.4 Enyl complexes

η^3-Allyl complexes are the simplest case of a series of open and closed polyenyl compounds:

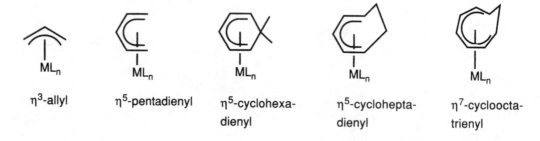

η^3-allyl

η^5-pentadienyl

η^5-cyclohexa-
dienyl

η^5-cyclohepta-
dienyl

η^7-cycloocta-
trienyl

Synthesis by electrophilic attack on neutral olefin complexes

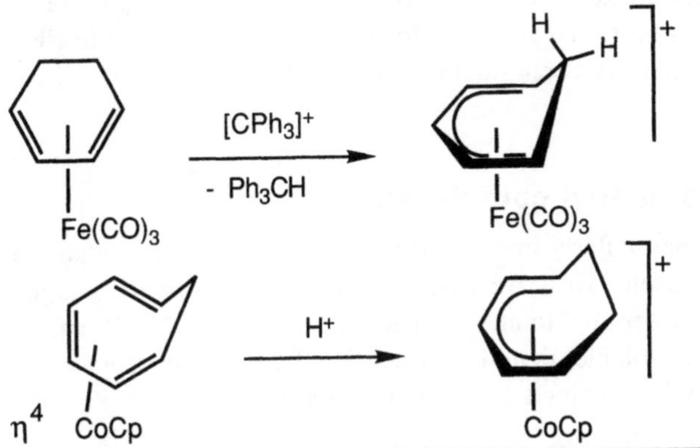

hydride abstraction

protonation

Synthesis by nucleophilic attack

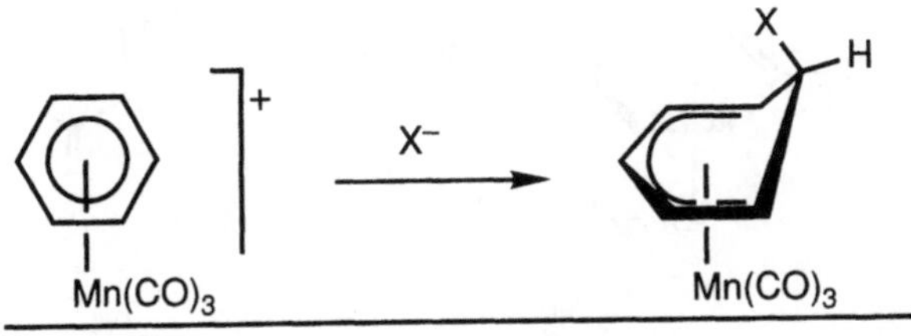

addition of nucleophiles
to cationic complexes;
X = H, alkyl

Synthesis by isomerization

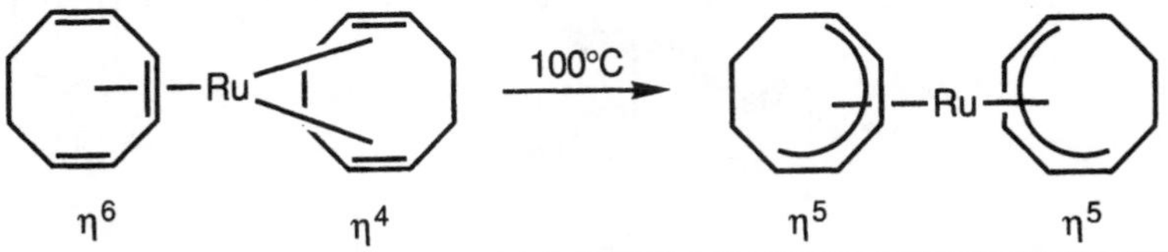

Synthesis from metal halides by anion exchange

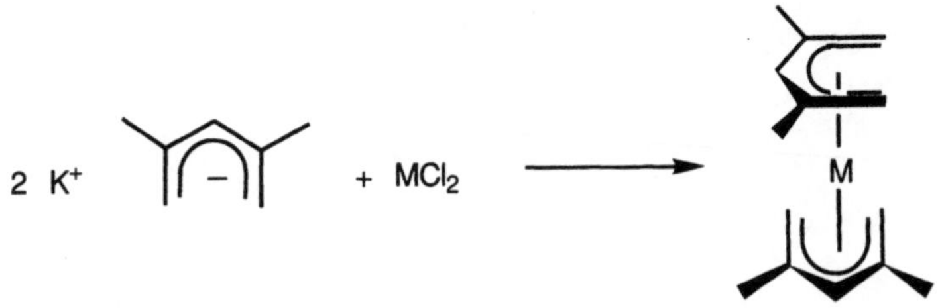

M = Ti, V, Cr, Mn, Fe

'Open sandwich' compounds of this kind provide a comparison with cyclopentadienyl complexes. Note that even Ti(dienyl)$_2$ (14 VE) is stable, whereas TiCp$_2$ is not. Ti(dienyl)$_2$ forms an adduct an with PF$_3$, Ti(dienyl)$_2$(PF$_3$), which has a bent sandwich structure.

5.5 Reactivity of dienyl complexes

Dienyl complexes, particularly iron tricarbonyl compounds, have important applications in stereoselective synthesis. Since they are positively charged, the complexes are susceptible to attack by a multitude of C-, N-, P- and O-nucleophiles to give isolable diene iron tricarbonyl complexes from which the free ligand may be obtained on oxidative degradation of the Fe(CO)$_3$ fragment.

6 Arene complexes

6.1 Bis-(arene)metal complexes

Following the recognition in the early 1950s of the sandwich bonding principle for metallocenes, based on the 6π-Hückel aromatic anion $C_5H_5^-$, the existence of related sandwich complexes of neutral arenes such as benzene was demonstrated by the synthesis of bis-(benzene)chromium, $Cr(C_6H_6)_2$. The complex is isoelectronic to ferrocene and follows the 18-electron rule:

E. O. Fischer (1955). In fact, Cr arene complexes had been prepared by F. Hein as early as 1919, although the true nature of these compounds was not discovered until 1954.

$$3 \ CrCl_3 + 2 \ Al + 6 \ C_6H_6 \quad \xrightarrow[\text{2) } H_2O]{\text{1) } AlCl_3} \quad \left[\underset{Cr}{\bigcirc} \right]^+ \quad \xrightarrow[\text{KOH}]{S_2O_4^{2-}} \quad \underset{Cr}{\bigcirc} \quad \begin{array}{c} 6\pi \\ d^6 \\ 6\pi \\ \hline 18 \ VE \end{array}$$

In this method a mixture of Al and $AlCl_3$ are used to reduce the Cr^{III} salt to zerovalent Cr. A modern alternative of synthesizing complexes of zerovalent metals is the co-condensation of atomic metal vapour under high vacuum with volatile ligands and solvent on the cold ($-196°C$) wall of the condensation reactor.

$$M \ (g) + C_6H_{6-n}R_n \quad \longrightarrow \quad \underset{Cr}{\overset{R_n}{\bigcirc}} \quad \begin{array}{l} M = Ti, \ Zr, \ Hf, \ Nb, \ Cr; \\ R = H, \ Me, \ Bu^t, \ etc. \end{array}$$

Bonding in bis-(arene)metal complexes resembles that in metallocenes, and the MO diagram is qualitatively very similar (cf. p. 47). Compared to free benzene, the C–C bonds in $Cr(C_6H_6)_2$ are slightly elongated; the ring–ring distance is almost equal to the van der Waals distance between two π-systems.

Since, unlike Cp, the aromatic ligands in bis-(arene) complexes are not negatively charged, there is no electrostatic contribution to bonding, and bis-(arene) complexes are in general less stable than metallocenes and more easily oxidized. Complexes with 16–21 VE exist:

16 VE	17 VE	18 VE
$Ti(C_6H_6)_2$ $V(C_6H_3Me_3)_2^+$	$[Ti(C_6H_5Ph)_2]^-$ $V(C_6H_6)_2$ $Cr(C_6H_6)_2^+$	$V(C_6H_6)_2^-$ $Cr(C_6H_6)_2$ $Fe(C_6Me_6)_2^{2+}$ orange, very stable

19 VE	20 VE	21 VE
$Fe(C_6Me_6)_2^+$ Deep purple	$Fe(C_6Me_6)_2$ Sensitive, isolable, possibly η^6,η^4	
$Co(C_6Me_6)_2^{2+}$ $\mu = 1.73$ B.M.	$Co(C_6Me_6)_2^+$ $\mu = 2.95$ B.M. Both rings are η^{6-} bonded	$Co(C_6Me_6)_2^0$ Isolable, thermally unstable, $\mu = 1.86$ B.M., possibly not η^6,η^6
	$Ni(C_6Me_6)_2^{2+}$ $\mu = 3.0$ B.M.	

Dicationic arene complexes such as $Fe(C_6Me_6)_2^{2+}$ are susceptible to nucleophilic attack. Their electron-attracting character is also revealed e.g. by the formation of charge-transfer complexes with electron-rich arenes such as diaminobenzenes and even with ferrocene:

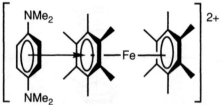

Appearances can be deceptive. Although reduction of $[Ru(C_6Me_6)_2]^{2+}$ gives $Ru(C_6Me_6)_2$, apparently a 20-electron species (comparable to the 20 VE cation $[Co(C_6Me_6)_2]^+$ and nickelocene), variable temperature NMR on this diamagnetic complex reveals that one of the arene rings is only η^4-bonded, so that the 18 VE configuration is maintained. This is another example of ligand fluxionality (cf. p. 60): the coordination modes of the arene rings interchange sufficiently fast so that at higher temperature both C_6Me_6 rings appear identical on the NMR time scale. In contrast to the Co analogue, $[Rh(C_6Me_6)_2]^+$ has an η^6,η^4 structure (18 VE); it is possible that this bonding mode is also adopted by $Fe(C_6Me_6)_2$.

$$\eta^6$$

$$d^8$$

$$\eta^4$$

6.2 Arene half-sandwich complexes

As is the case with Cp compounds, arene complexes of half-sandwich structure, especially those with carbonyl ligands, have proved particularly valuable synthetically.

Arene half-sandwich complexes have an interesting history. It had been known for a long time that Ag^+ forms adducts with arenes even in aqueous solution, although the structure of these adducts was unknown. As early as 1949 Andrews and Keefer proposed an η^6-benzene coordination to Ag^+; and even suggested a tripledecker structure for the binuclear dication, $(Ag^+)_2(C_6H_6)$ —two years before the discovery of the ferrocene structure and the sandwich-bonding principle. In reality, however, benzene interacts with Ag^+ in η^2-fashion.

Synthesis from metal carbonyls

yellow, air-stable

Synthesis by ligand exchange

Synthesis from dienes

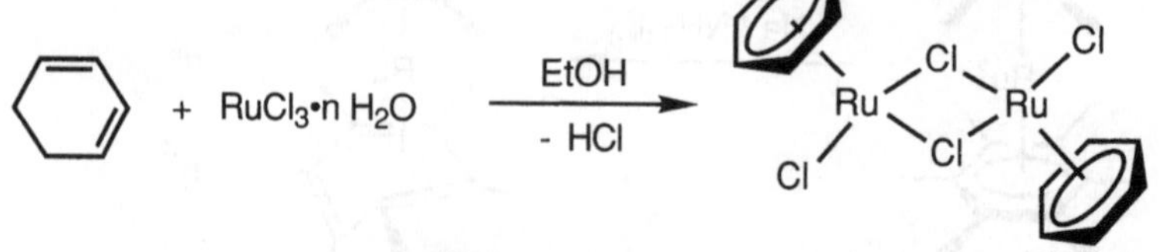

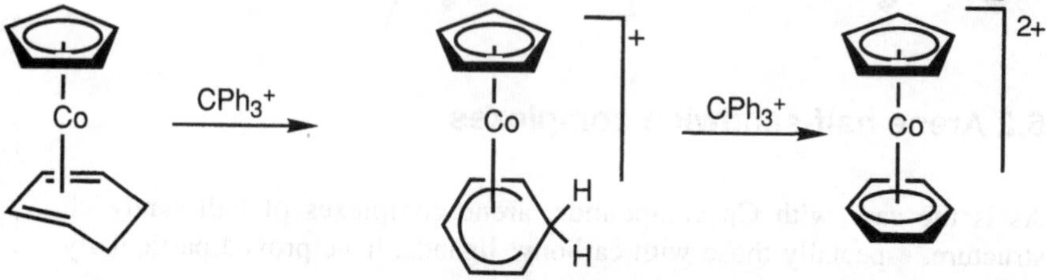

Two metals may be coordinated to the same side of an arene to give Dewar-benzene like structures:

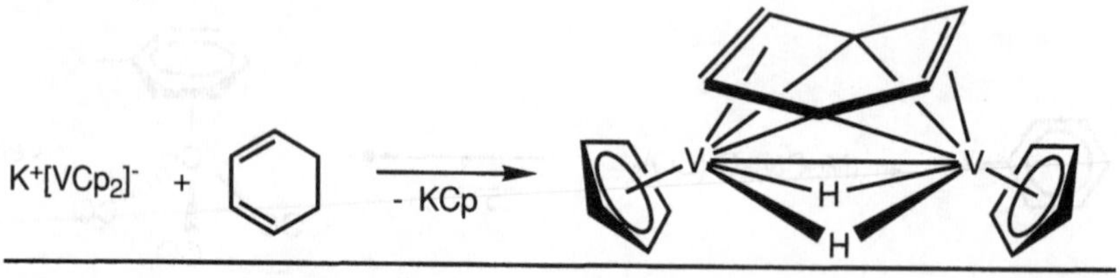

Synthesis from alkynes

Alkynes may be cyclotrimerized to arenes (cf. pp. 35, 37). Mononuclear or binuclear arene complexes may result. The structure of the binuclear iron complex below is closely related to the vanadium complex above.

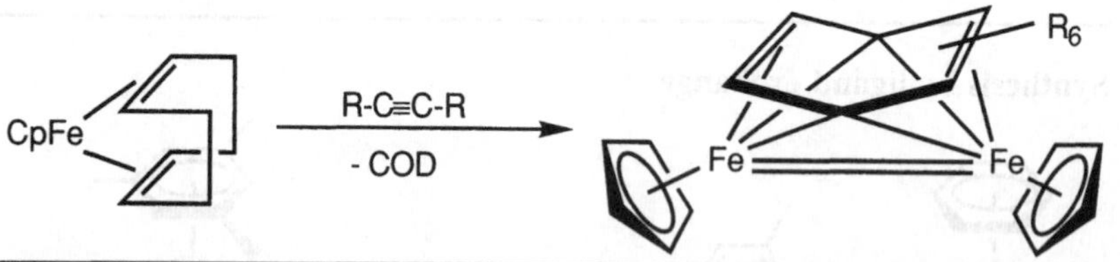

Synthesis from metal halides

Arenes form π-complexes with strong Lewis acids such as some metal halides. Examples are the yellow solutions of $TiCl_4$, $ZrCl_4$, $NbCl_5$ etc. in benzene or toluene. With electron-rich arenes stable complexes may be isolated:

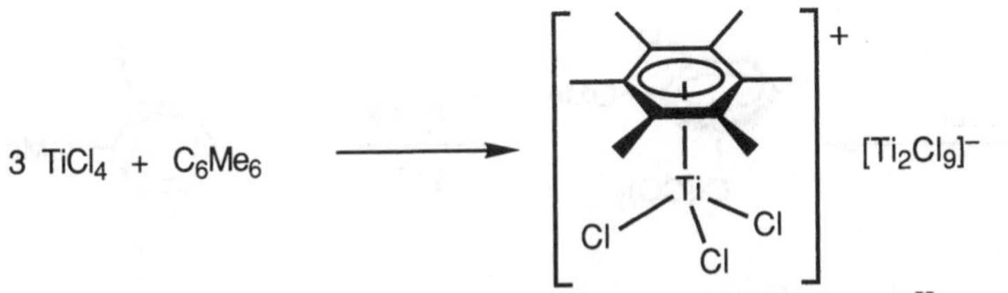

$$3 \ TiCl_4 \ + \ C_6Me_6 \ \longrightarrow \ \left[\ \underset{Cl}{\overset{\displaystyle Ti}{}} \ \right]^{+} \ [Ti_2Cl_9]^{-}$$

The reduction of $TiCl_4$ in the presence of $AlCl_3$ gives Ti^{II} complexes which are polymerization catalysts on addition of aluminium alkyls:

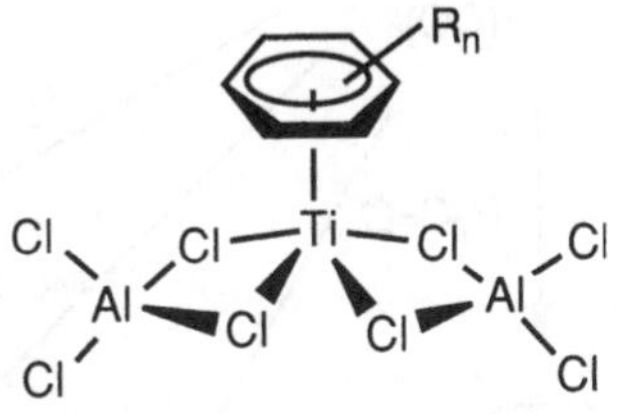

Reactivity

The changes in the reactivity pattern of coordinated arenes in half-sandwich complexes is best illustrated for arene chromium tricarbonyl complexes, synthetically the most important group in this class. The $Cr(CO)_3$ fragment is strongly electron-withdrawing and activates the arene ligand towards **nucleophilic attack**. This activation is most apparent in the ease of nucleophilic aromatic substitution of aryl halides, which for free arenes requires very drastic reaction conditions.

With alkyl substituted arenes, coordination to $Cr(CO)_3$ greatly enhances the acidity of α-alkyl protons.

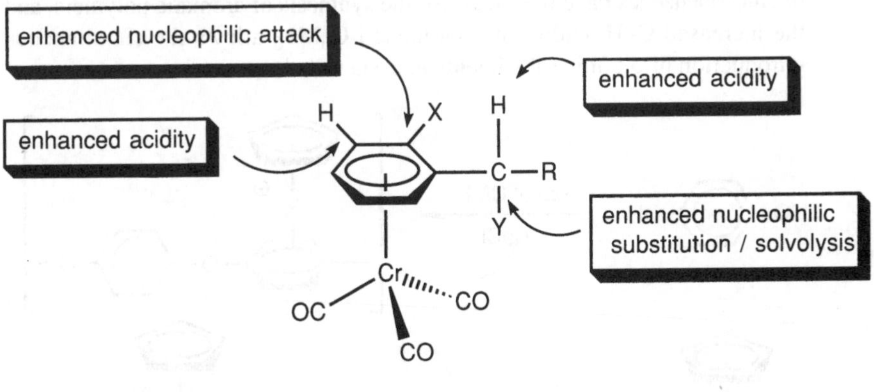

These properties can be exploited in numerous ways. For example, the introduction of alkyl groups can be 100% regioselective:

Other metal fragments have similar electronic properties to $Cr(CO)_3$, particularly if they carry a positive charge. For example, $CpRu^+$ complexes of chlorobenzenes have been used for the synthesis of aromatic polymers, and the increased C–H acidity of coordinated C_6Me_6 allows the simultaneous introduction of olefinic substituents in a 'starburst' reaction:

6.3 η² to η⁴ coordinate arenes

Although η^6 sandwich-type coordination is most common for arenes, other coordination modes may be adopted. An example for η^2-benzene is [Os(η^2-C$_6$H$_6$)(NH$_3$)$_5$]$^{2+}$. In this kinetically stable complex the coordinated C=C bond is protected while the non-coordinated bonds may be selectively hydrogenated. Ligand exchange becomes possible after oxidation of Os(II) (d^6) to Os(III) (d^5) so that a stepwise cyclic process can be devised:

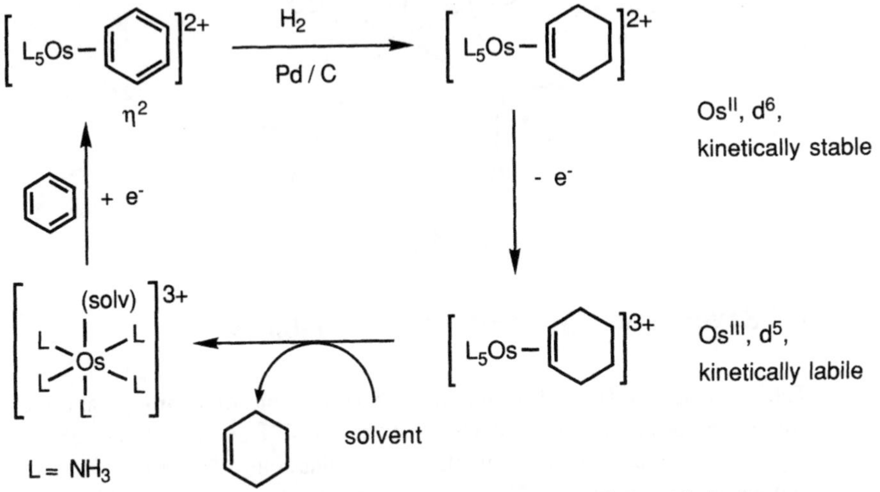

Nickel η^2-benzene complexes are also known; reaction with hydrogen gives a Ni(I) hydride:

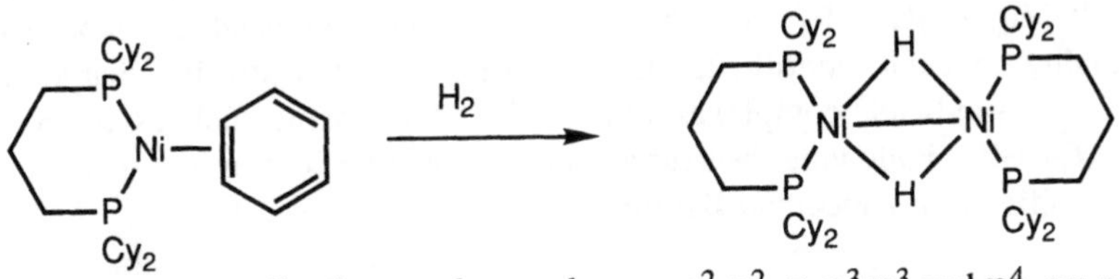

Other arene coordination modes, such as μ-η^2:η^2, μ-η^3:η^3 and η^4, are a reflection of the ability of arene ligands to accommodate the electronic requirements of the metal:

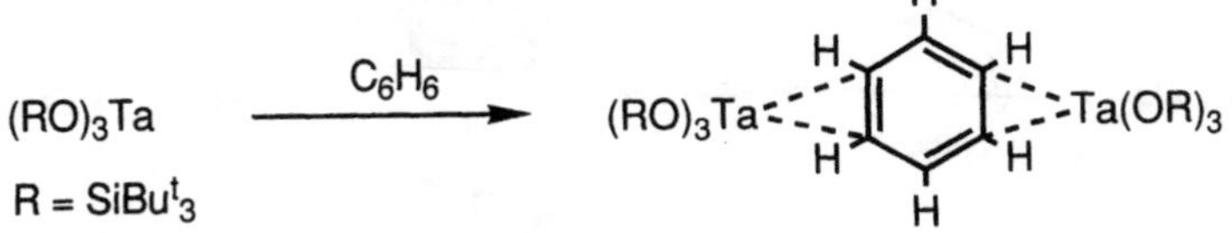

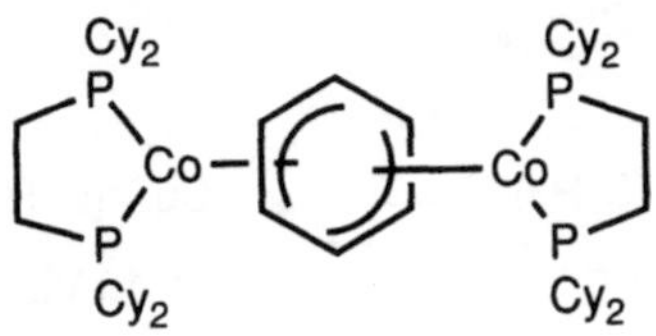

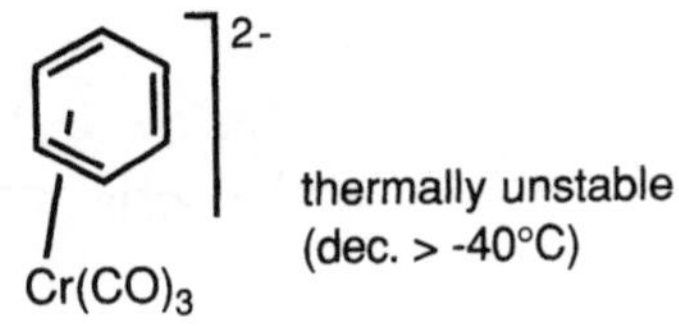

An interesting bonding mode is displayed by benzene coordinated to the M_3 unit in some trinuclear metal carbonyl clusters, such as $Os_3(C_6H_6)(CO)_9$. The ligand is attached to the three Os atoms via three localized C=C bonds which are significantly shorter than the other three bonds in the ring. A similar type of bonding is being discussed for benzene chemisorbed onto the surfaces of heterogeneous metal catalysts.

$$\mu_3\text{-}\eta^2\text{:}\eta^2\text{:}\eta^2\text{-}C_6H_6$$

6.4 Seven- and eight-membered ring ligands

The sequence of Hückel aromatic systems capable of forming sandwich complexes is continued with $C_7H_7^+$ (6π) and $C_8H_8^{2-}$ (10π). The larger $C_8H_8^{2-}$ ring is particularly suitable for coordination to larger metal centres and with U(IV) gives $U(C_8H_8)_2$, 'uranocene'. In contrast to Cp and benzene, changes in hapticity (η^1 to η^n-bonding) are frequently encountered, depending on the electron count of the metal. $C_7H_7^+$, the *tropylium cation*, forms stable salts, e.g. $[C_7H_7]BF_4$. Even though bonding in organometallic complexes is highly covalent, coordinated $C_7H_7^+$ carries a partial positive charge, and mixed sandwich complexes have a dipole moment, such as $(C_5H_5)^{\delta-}V(C_7H_7)^{\delta+}$. Following the neutral electron counting convention, $\eta^7\text{-}C_7H_7$ is classified as a **7-electron ligand**.

Preparation of C_7H_7 complexes by ligand substitition

$$CpV(CO)_4 \; + \quad \longrightarrow \quad + \; 1/2\,H_2 \; + \; 4\,CO$$

$$CpCr(C_6H_6) + C_7H_8 \xrightarrow{AlCl_3} [CpCr(C_7H_7)]^+ \xrightarrow{e^-} CpCr(C_7H_7)$$
$$\qquad\qquad\qquad\qquad\qquad\qquad\quad \text{17 VE} \qquad\qquad\quad \text{18 VE}$$

Preparation from cycloheptatriene complexes

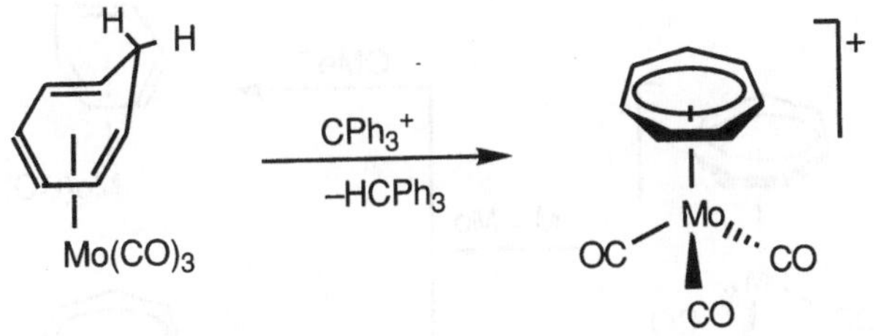

hydride abstraction

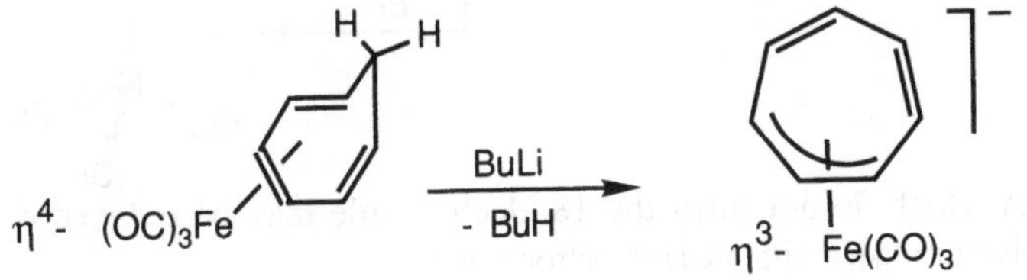

proton abstraction

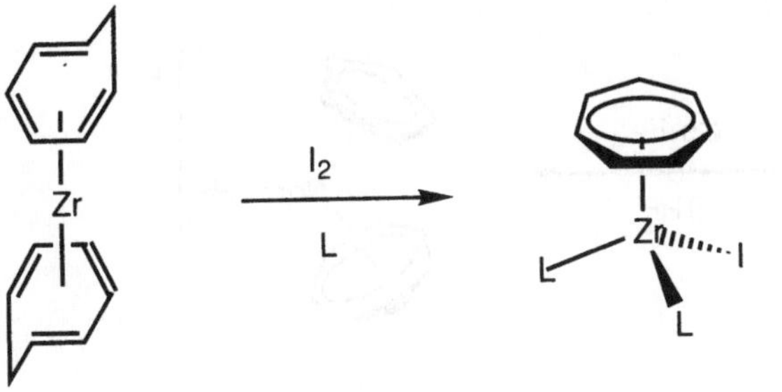

L = THF, PMe$_3$

Preparation by reduction

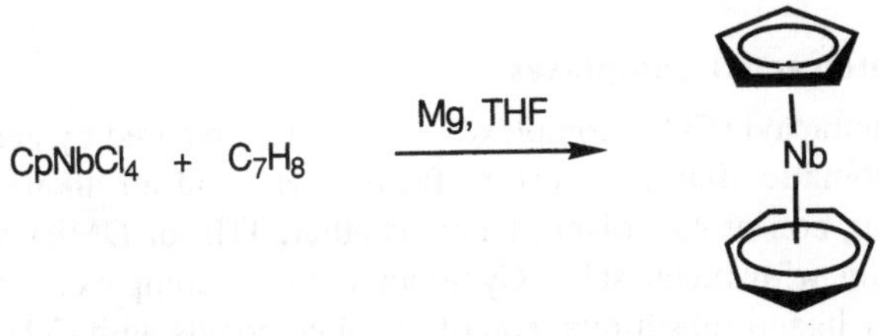

$$CpNbCl_4 \ + \ C_7H_8 \xrightarrow{\text{Mg, THF}}$$

Reactions of cycloheptatrienyl complexes

Nucleophilic attack can occur on the C$_7$H$_7$ ring or on the metal, depending on the nucleophile and the metal centre.

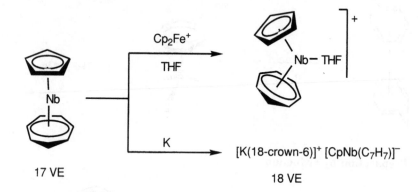

Complexes which do not fulfil the 18-electron rule may be reduced or oxidized to give products with an 18-electron count:

17 VE   18 VE

Cyclooctatetraenyl complexes

Cyclooctatetraene is a yellow, non-conjugated olefin (8π!); it exists in chair and tub conformations. The air-sensitive dianion $C_8H_8^{2-}$ by contrast is planar (10π) and deeply coloured.

Cyclooctatetraenyl (COT) complexes are usually prepared by generating the planar, aromatic dianion $C_8H_8^{2-}$ from C_8H_8 and an alkali metal in a coordinating non-protic solvent (such as ether, THF or DME) and reacting this solution with metal salts. Cyclooctatetraene complexes may also be obtained by ligand substitution e.g. of metal carbonyls with C_8H_8. The COT ligand may adopt planar or non-planar conformations. Non-planar η^4 complexes are usually fluxional and may be conjugated or non-conjugated:

M = U ← $MCl_4 + 2\ K_2C_8H_8$ → M = Zr

Uranocene

$$CpCo(CO)_2 + C_8H_8 \longrightarrow$$

tub conformation,	chair conformation,
non-conjugated,	conjugated,
$1,2\eta{:}5,6\eta$, rigid	$1\text{-}4\eta$, fluxional

In binuclear COT complexes, too, the bridging ligand may adopt a number of different conformations, with the metal bound to the same or to opposite faces of the ring. In the electron-deficient compound $Ti_2(COT)_3$ each Ti interacts with one planar terminal η^8-ring and four carbons of the bridging C_8H_8. Electrochemical reduction gives the highly sensitive tripledecker anion $[Ti_2(COT)_3]^{2-}$; the additional electrons are delocalized over the central ring which now forms a planar aromatic system.

$$CrCl_2 + 2\ NaCp + K_2C_8H_8 \xrightarrow[\text{2) }150°C]{\text{1) RT}}$$

$$Ti(OBu)_4 + C_8H_8 + AlEt_3 \quad 2:4:20$$

$\mu\text{-}\eta^4{:}\eta^4$

$\xrightarrow{e^-}$

$\mu\text{-}\eta^8{:}\eta^8$

6.5 Heteroarene complexes

The presence of electron pairs on the heteroatom means that these molecules can act not only as π- but also as n-donors; in some, this bonding mode predominates, e.g. with pyridine where it has only recently been possible to isolate π-complexes such as $Cr(\eta^6\text{-}C_5H_5N)_2$.

Arenes containing heteroatoms in their rings are frequently able to form π-complexes similar to those described in Chapters 4 and 6. A large number of potential ligands are available, not all of which are known in the free state. The number of electrons these ligands contribute to the metal varies; the substitution of a ring-carbon atom by boron for example reduces the electron count by one. Some examples:

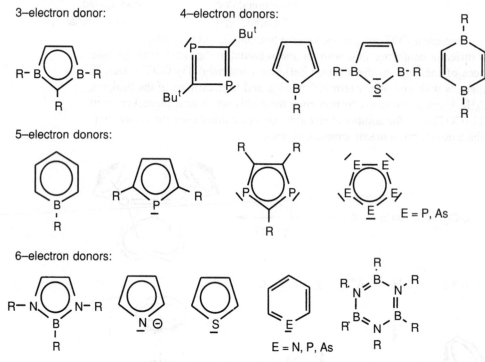

3–electron donor:

4–electron donors:

5–electron donors:

E = P, As

6–electron donors:

E = N, P, As

Most heteroarene complexes obey the 18-electron rule. N and P heteroatoms possess electron pairs oriented perpendicular to the π system which may act as n-donors.

The chemistry of sulphur heterocycles has attracted attention since these complexes shed light on the process of sulphur removal from petrochemical feedstocks (*'hydrodesulphurization'*). Numerous thiophene complexes are known; $Ru(C_4H_4S)_2$ is fluxional.

A special case of heterocycles are the carboranes. The most prominant example is the 'dicarbollide' dianion, $[C_2B_9H_{11}]^{2-}$ (M. F. Hawthorne, 1967). It has bonding properties similar to the cyclopentadienyl ligand, and a large number of metallocene analogues are known, e.g.:

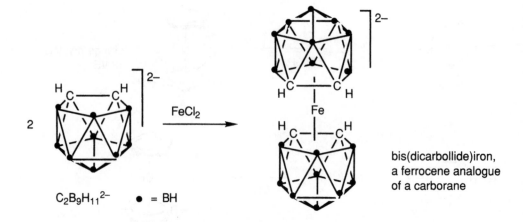

$C_2B_9H_{11}^{2-}$ ● = BH

bis(dicarbollide)iron,
a ferrocene analogue
of a carborane

6.6 Multidecker complexes

The Hückel theory predicts a particularly stable electron configuration for planar π systems if $(4n+2)\pi$ electrons are present, i.e. 6 π electrons in the case of benzene. When two such ligands are coordinated to a metal, the most stable configuration is reached if the number of electrons involved in bonding amounts to 18. Continuing the stacking process can be expected to lead to multidecker sandwich complexes with a series of characteristic electron configurations, so-called 'magic numbers' which indicate energy minima for multidecker structures: 18, 30, 42, 54, etc.

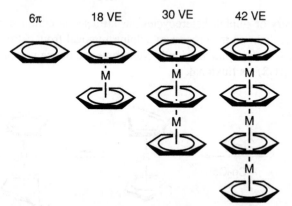

Such multidecker complexes do indeed exist. Most tripledeckers, for example, have 30 VE. Different numbers of electrons can however be accommodated, and systems with 26 to 34 VE are known. The tripledecker $[Ni_2Cp_3]^+$ (34 VE) has already been mentioned (p. 54). Co-condensation of an excess of chromium atoms with arenes gives $Cr_2(arene)_3$ (30 VE).

Cr(g) + [mesitylene] →(co–condensation)→ [bis(mesitylene)chromium multidecker] 30 VE

Boron ligands have proved particularly successful in multidecker synthesis. Tetra-, penta-, hexa- and even polydecker complexes are known (W. Siebert, 1985).

It is advantageous for the synthesis of multidecker complexes if either the metal fragment, e.g. CpM, or the bridging ligand is electron-deficient; as with the 26 VE vanadium complex and the successful use of boron heterocycles in multidecker synthesis.

CpV(allyl)$_2$ + [cyclohexadiene] →(100°C)→ [V multidecker] 26 VE

[Cp–Fe–C$_6$H$_6$]$^+$ + RuCp*$_2$ →(hv, –C$_6$H$_6$)→ [Fe–Ru multidecker]$^+$ 30 VE

[boron heterocycle with Me, H] + CpM(CO)$_x$ →(Δ)→ [multidecker product]

M = Fe, Co and/or Ni

MM' = FeFe, FeCo, CoCo, CoNi, NiNi, NiNi$^-$ (29–34 VE)

[polydecker Co/Ni/Co complex] 70 VE

Further reading

A number of excellent textbooks will provide additional coverage of the material discussed here, for example:

Cotton F.A. and Wilkinson G. (1986). *Advanced Inorganic Chemistry*, (5th edn), Wiley, Chichester.

Elschenbroich C. and Salzer A. (1992). *Organometallics: a concise introduction*, (2nd edn), VCH Weinheim.

For a comprehensive treatise on all aspects of organometallic chemistry see:

Wilkinson G., Stone F. G. A. and Abel E. W. (eds) *Comprehensive organometallic chemistry*, (1982). 9 volumes, Pergamon Press, Oxford.

Preparative techniques are described in:

Shriver D. F. and Drezdon M. A. (eds) *The manipulation of air-sensitive compounds,,* (1986). 2nd edn. Wiley, Chichester.

Wayda A. L. and Darensbourg . Y. (eds), (1987). *Experimental organometallic chemistry*, ACS Symposium Series 357, American Chemical Society.

Books and articles on specialized topics:

A collection of reviews an several aspects of organometallic chemistry has appeared in : *Chemical Reviews*, (1988), **88**, 989–1421.

Poli R. (1991)Monocyclopentadienyl halide complexes of d– and f–block elements. *Chemical Reviews*, **91**, 509.

Winter M. J. (1989). Unsaturated dimetal cyclopentadienyl carbonyl complexes. *Advances in Organometallic Chemistry*, **29**, 101.

Wilke G. (1988). Contributions to the organic chemistry of nickel, *Angewande Chemie International Edition in English*, **27**, 185.

Herrmann W. A. (1988). Organometallic chemistry in high oxidation states: the example of rhenium, *Angewande Chemie international Edition in English*, **27**, 1297.

Yasuda H. and Nakamura A. (1987), Diene, alkyne, alkene and alkyl complexes of early transition metals: structures and synthetic applications in organic and polymer chemistry, *Angewande Chemie International Edition in English*, **26**, 723.

Jolley P. W. (1985). η^3–Allyl Palladium Complexes, *Angewande Chemie International Edition in English*, 1985, **24**, 283.

Siebert W. (1985). Diborole complexes with activated C–H bonds: building blocks for multidecker sandwich complexes. *Angewande Chemie International Edition in English*, **24**, 943.

McQuillan F. J., D.G. Parker and G.R. Stephenson, (1991) *Transition metal organometallics for organic synthesis,* Cambridge University Press.

Index